The OAU After Twenty Years

Also of Interest

African Security Issues: Sovereignty, Stability, and Solidarity, edited by Bruce E. Arlinghaus

Crisis in Africa: Battleground of East and West, Arthur Gavshon

Military Development in Africa: The Political and Economic Risks of Arms Transfers, Bruce E. Arlinghaus

State Versus Ethnic Claims: African Policy Dilemmas, edited by Donald Rothchild and Victor A. Olorunsola

African International Relations: An Annotated Bibliography, Mark W. DeLancey

Ethnicity in Modern Africa, edited by Brian M. du Toit

Regionalism Reconsidered: The Economic Commission for Africa, Isebill V. Gruhn

African Upheavals Since Independence, Grace Stuart Ibingira

Africa's International Relations: The Diplomacy of Dependency and Change, Ali A. Mazrui

Alternative Futures for Africa, edited by Timothy M. Shaw

Africa and the West, edited by Michael A. Samuels

Nonstate Actors in International Politics: From Transregional to Substate Organizations, Phillip Taylor

*Available in hardcover and paperback.

Westview Special Studies on Africa

The OAU After Twenty Years
Amadu Sesay, Olusola Ojo, and Orobola Fasehun

This book investigates the performance of the Organization of African Unity (OAU) since its inception, focusing on four areas of central concern to African states: decolonization, conflict control, development, and human rights. The authors examine the OAU's record against the challenge of apartheid and the OAU's lack of resources and effective sanctions. They make a number of suggestions for enhancing the OAU's future viability and its ability to address the continent's pressing economic and social needs.

Dr. Amadu Sesay is a lecturer in the Department of International Relations at the University of Ife, Nigeria. He is coeditor (with Ralph Onwuka) of *The Future of Regionalism in Africa*. *Dr. Olusola Ojo* is senior lecturer in the Department of International Relations at Ife. He is author of *Africa and the Arab World* and coeditor (with Timothy Shaw) of *Africa and the International Political System*. *Dr. Orobola Fasehun* was a lecturer in the Department of International Relations at Ife and has published many articles on Nigerian foreign policy and on the OAU.

The OAU After Twenty Years

Amadu Sesay, Olusola Ojo, and
Orobola Fasehun

Westview Press / Boulder and London

The paper used in this publication meets the minimum
requirements of the American National Standard for
Permanence of Paper for Printed Library Materials
Z39.48-1984.

Westview Special Studies on Africa

This is a Westview softcover edition, manufactured on our own premises using
equipment and methods that allow us to keep even specialized books in stock.
It is printed on acid-free paper and bound in softcovers that carry the
highest rating of NASTA in consultation with the AAP and the BMI.

Published in 1984 in the United States of America by Westview Press, Inc.,
5500 Central Avenue, Boulder, Colorado 80301; Frederick A. Praeger,
Publisher

Library of Congress Catalog Card Number: 84-52149
ISBN: 0-8133-0112-2

Composition for this book was provided by the authors
Printed and bound in the United States of America

10 9 8 7 6 5 4 3 2 1

Contents

Preface

The year 1983 marked the end of the second decade of the Organisation of African Unity (OAU). This volume is an assessment of the performance of the Organisation since its creation in 1963 and of its potential, focusing on Africa's central concerns: decolonisation, African conflicts, economic development, and human rights.

Three main factors inform the study. First, the OAU is Africa's largest and most prestigious regional organisation and, as such, offers a broad view of the various foreign policies of African states. The difficulties as well as the latent potential of the Organisation have attracted the attention of both African and non-African scholars. This leads to the second reason, which is that the OAU, being a regional organisation, offers a laboratory for the testing of neo-functionalist claims that regional organisations are more effective in dealing with regional problems than are global international organisations. Third, the study is an addition to the literature on Africa's international relations in particular and regional organisations in general.

In the course of preparing the book, we received assistance from many individuals and institutions. We are particularly grateful to the University of Ife for a research grant that enabled us to visit Abidjan, Addis Ababa, and Dar-es-Salaam. We are also grateful to those individuals in the African Development Bank, the OAU Secretariat, the OAU Liberation Committee Headquarters, and the Economic Commission for Africa who provided us with invaluable information and documents but who preferred to remain anonymous for obvious reasons. Last but by no means the least, we wish to express our gratitude to all our families for lovingly coping with our erratic working schedules in the course of the book's preparation.

Amadu Sesay
Olusola Ojo
Orobola Fasehun
Ife, Nigeria

1
Introduction

The roots of the Organisation of African Unity (OAU), which was set up in 1963 at a summit meeting of thirty-two African Heads of States and Governments, can be traced back to a major discourse that started in the penultimate days of colonialism on the continent. The debate focused on both the nature and character of future relationships among the independent African states and was expressed within the Pan-African Movement.[1]

There were two broad and conflicting schools of thought. The first within this grouping--the "radical school"--advocated immediate political union among the independent states. The second--"the moderates cum-conservatives"--was opposed to political union and called instead for close cooperation in other (i.e., economic, social, cultural, and educational) fields. These conflicting positions on the nature and form that African unity was to take led to the formation of various ideological groupings between 1960 and 1963, reflecting the radical-moderate split within the Pan-African Movement. The most important blocs--institutional expressions of these divisions--were the Brazzaville, the Monrovia, and the Casablanca.[2] Besides these ideological divisions on the broad issues of unity among the independent states, there were other specific issues like the Congo crisis, the Algerian war of independence, and the status of Mauritania, which sharpened the prevailing differences among the independent African states.

Despite these differences, diplomatic moves were made during this period by the "uncommitted" or "neutral" African states, most notably Ethiopia, to bring about a reconciliation of the different ideological blocs within the Pan-African Movement. Their main objective was to call a conference of all independent states with a view to creating a Pan-African organisation that would accommodate and moderate the disparate positions on African unity.

THE ADDIS ABABA CONFERENCE, MAY 1963

A number of propitious factors helped to make
this conciliatory diplomatic initiative by the neutrals a
success. First, with the formal independence of Algeria
in 1962, the question of whether Algerian nationalists
should be seated at pan-continental meetings (a very
divisive issue in the past) was effectively resolved.
Second, there was a lull in the tensions and conflict in
the Congo following the assassination of Patrice Lumumba
in 1961. Although some African states remained hostile
to the new authorities in that country, by the end of
1962 such opposition was not as strong as it had been
while Lumumba was alive. And third, it was becoming
apparent that the factional politics that had plagued
Pan-Africanism since the late 1950s could not be allowed
to continue to debilitate the Movement much further.
Many African leaders had come to think that it was in
Africa's interest to meet and resolve outstanding
differences once and for all. Thus, the way was paved
for the May 1963 African heads of state summit in Addis
Ababa, Ethiopia.

The summit itself was preceded by a meeting of
Foreign Ministers on May 15. These ministers grouped
themselves into two committees: a political committee and
another charged with preparing a draft charter. After
much deliberation, they made the following recommendations for the heads of state to consider: (1) machinery
should be established to prepare a "Master Charter" to
which all African states would subscribe under a continental organisation; (2) money and material support should
be provided to the liberation movements; (3) there should
be concerted efforts among the independent states in the
areas of economics, health, science and culture; and
(4) the heads of states should look into the possibility
of establishing an African Common Market.

Emperor Haile Selassie opened the subsequent summit
on May 22 with a call to the heads of state to eliminate
colonialism and white minority governments from the
continent. On the crucial and controversial issue of
unity, the Ethiopian monarch opined that "the union we
seek can only come gradually as day-to-day progress which
we achieve carries us slowly but inexorably along this
course."[3] He appealed to his colleagues not to delay the
adoption of a charter. The African leaders then took
turns professing their commitment to African unity and
chorused the litany of opposition to colonialism. So, on
23 May 1963, the Charter of the Organisation of African
Unity was duly signed by the assembled African leaders.

THE CHARTER OF THE OAU

The Charter of the OAU reflected the victory of the
moderates over the radicals in the debate about the
nature and form of post-independence inter-state relations on the continent. The Charter does not depart
radically from traditional intergovernmental institutions
such as the United Nations (UN), whose Charter had a
pervasive influence on the African leadership. For
instance, in the Preamble of the OAU Charter African
leaders state that they are "persuaded that the Charter
of the United Nations and the Universal Declaration of
Human Rights . . . provided a solid foundation for peaceful and positive cooperation among states."
The OAU Charter contains the various collective
concerns of the African leaders: continental decolonisation, security problems and economic development. The
Preamble acknowledges the need to promote understanding
among the African peoples as well as cooperation among the
independent states. To achieve these objectives, the
African leaders recognised that "conditions for peace and
security must be established and maintained" within the
continent. In that regard, they expressed their
determination "to safeguard and consolidate our hard-
won independence as well as the sovereignty and territorial integrity of our states, and to fight against
neocolonialism in all its forms." An awareness of the
negative impact of the forces of disunity among their
ranks, particularly since 1960, caused African leaders
explicitly to express the desire "that all African states
should henceforth unite" and "to decide to reinforce the
links between our states by establishing and strengthening common institutions."

Objectives of the OAU

All international governmental institutions preface
their constitutions with a general statement of purposes.
These purposes reflect the broad consensus of the
constituent members and the ideal goals of the organisation. In this sense, the OAU is a characteristic inter-
state institution. However, the content of the preamble
is rather distinctive in this case. There are three
broad identifiable purposes in the OAU Charter. First
the primary goal is the promotion of the unity and
solidarity of the continent. This purpose was inserted
in order to continue the search for unity which had
eluded African states since the first wave of independence. By making African unity a goal that is to be

pursued continuously member states of the Organisation
sought to enhance their resistance to neo-colonial
manipulation of interstate differences, i.e. to combine
sovereignty with security. Second, unity is closely
linked with another core purpose: defense of sovereignty,
territorial integrity and independence. Given the
fragility of many African states and the disintegrative
trend which had so soon manifested itself in the former
Congo (now Zaire) the insertion of this purpose is clearly
understandable. Another reason for inserting this clause
is to forestall irredentia, subversion and future
imperialist designs by the bigger African states.
Finally, the defence of sovereignty and territorial
integrity of member states was intended to pre-empt
foreign intervention in Africa from outside the continent.

Third, a final broad purpose was the eradication of
"all forms of colonialism from Africa." There are many
reasons for the inclusion of this goal. For many African
states, security is seen to be indivisible. Moreover,
the unity, solidarity, and territorial integrity of member
states will be difficult to achieve whilst large and
strategic sections of the continent remain under colonial
or white minority regimes. Besides, colonialism sub-
jugated Africans and created in them a feeling of
inferiority. For many Africans, therefore, anti-
colonialism was seen to be a moral crusade and a pre-
requisite for the realisation of the African personality,
let alone the *sine qua non* for development.

Principles of the OAU

A statement of the purposes of the OAU Charter is
followed by the principles in Article III. The principles set out the guidelines for interstate conduct and
fortify the purposes stated in Article II.

The most fundamental principle is the sovereign
equality of all member states--a typical claim of international organisations and laws. The equality referred
to is that of legal status; it does not refer to size,
power and resources. Indeed the OAU, like the UN, tacitly
recognises inequality in assessing members' contributions
to the annual budget. Thus while Nigeria was asked for
6.99 percent of the budget or US$153,290.70 in 1976,
Central African Republic was assessed at 1.47 percent or
$32,237.10. Besides, the OAU has often selected Nigeria
to serve on many mediation panels. This is a recognition
by the Organisation of Nigeria's influence within the
continent as well as its importance in world affairs,
despite Nigeria's formal equality with all other members.

Three other principles--(1) non-interference in the
internal affairs of states; (2) respect for the
sovereignty and territorial integrity of members; and

3) condemnation of subversion activities--are interrelated
and are meant to reinforce each other. These principles
reflect the concerns of the moderate states over the
alleged promotion and support of subversive activities by
more radical states. In actual fact, member states have
not consistently adhered to these principles. Two
examples will suffice. First, Ghana under Nkrumah was
accused of funding the efforts to overthrow violently
moderate regimes, particularly the government of Houphouet-
Boigny in the Ivory Coast. And second, Libya has also
been accused by Gambia of masterminding the abortive
July 1981 coup.

Many of the above principles could not be realised
in a situation of conflictful ambiance. To preempt as
well as to manage conflicts, the OAU articulated rules of
conduct to govern interstate differences. Member states
are enjoined to settle their disputes peacefully by
internationally sanctioned procedures; i.e. by negotiation, mediation, conciliation and arbitration. However,
as with all the other principles member states have
occasionally ignored this one. Thus, Ethiopia and
Somalia went to war in 1977 over disputed Ogaden.
Similarly, Algeria and Morocco fought over the Tindouf
region in 1963-64.

The list of principles concludes with the affirma-
tion of a policy of non-alignment with regard to all
ideological blocs. This principle is an attempt to prevent a spillover of the East-West cold war conflict into
the continent. Nonalignment is, however, merely
rhetorical. For many African states, including the
seemingly radical socialist ones, borrow developmental
paths that are either capitalist or socialist. Besides,
nearly all African states have close trade, aid, cultural
and technical links with their former metropoles. These
have compromised their nonalignment in practice.

From the above introduction to features and
principles of the Charter it is obvious that the rules
enunciated by African states are quite broad. Indeed,
they were deliberately made so in order to appeal to the
spectrum of ideological factions within the Pan-African
movement. We shall, in the rest of the book, examine in
greater detail the impact of these Charter provisions on
the ability of the OAU to respond to the demands of its
members since it was created in 1963.

STRUCTURE AND ORGANS OF THE OAU

In order to accomplish the objectives of the
Organisation, four major institutions were created. These
are the Assembly of Heads of State and Government, the
Council of Ministers, the General Secretariat, and the
Commission of Mediation, Conciliation and Arbitration.

The Assembly of African Heads of State and Government

At the apex of the OAU is the Assembly of African Heads of State and Government (AHG), which is composed of African Presidents, and Prime Ministers or their representatives. This is the supreme supervisory and decision-making organ of the Organisation. Its annual summits offer the opportunity for debating practically all issues of relevance to Africa constrained only by Charter limitations. The AHG is the most democratic and open of all Pan-African institutions; its members are not only equal--each has just one vote--they are also allowed to say whatever they feel during the summit meetings.

The authority of the OAU Assembly is on paper rather wide. The Charter states that the Assembly has the right to "discuss matters of common concern to Africa with a view to coordinating and harmonising the general policy of the Organisation." So, at first sight, it appears that the AHG, like the United Nations General Assembly, has the power to discuss the numerous and comprehensive issues touched upon by the Charter. This is, however, hardly the case, because of a basic legal and political contradiction: the power to "discuss matters of common concern to Africa" is hampered by a contrary Charter prohibition, namely "non-interference in the internal affairs of states." This means that any member state may prevent the discussion of any issue that may embarrass it. In practice, this provision has severely limited the Assembly's power. Besides, the OAU Charter does not have the equivalent of Chapter 7 of the UN Charter which empowers the world body to discuss issues which are perceived as threats to peace and security. A second and perhaps more important limitation on the powers of the AHG is the fact that its decisions are not enforceable. Unlike the UN, the OAU Charter does not contain any sanctions provisions. Consequently, its decisions are mere recommendations which member states may choose to obey or ignore at will.

Decision-making in the Assembly is by vote, although much is made of consensual solution to problems. The resolutions of the Assembly are "determined by a two-thirds majority of the members of the Organisation." However, procedural matters are exempt; they require only a simple majority. To reflect its "Africanness," annual meetings of the OAU are held in various African capitals. Emergency meetings are held in Addis Ababa upon the request of any member state and if such request is approved by two-thirds of the total membership of the OAU.[4]

Though not stated in the Charter itself--a virtually unchangeable document--the Assembly has institutionalized certain practices which have become part of its political procedure. For instance, there has emerged the practice of making the host head of state the chairman of the OAU until the following summit. This position of chairman is, however, titular, lacking in any putative executive authority. Nonetheless, the prestige of being Africa's spokes person for a year has lured many heads of mini-states to seek to host the OAU's annual conference. In the past, some African states opposed to particular regimes have tried to prevent the leaders of such regimes from hosting the summit and thus becoming chair of the Organisation. For example, some states opposed to former Ugandan leader, Idi Amin, did try to block his hosting of the summit in 1975 and thus his subsequent elevation to the OAU chairmanship. However, these states, led by Tanzania and strongly supported by Zambia and Mozambique, failed in their effort to prevent Amin from being OAU Chairman for 1975-76. This failure of Tanzania stems not so much from OAU love for Idi Amin but rather from the Organisation's dogmatic adherence to convention.[5]

When the Assembly meets, it performs one or more of the functions assigned to it by the Charter. The most comprehensive of these is its review function. The AHG reviews the structure and functions of any organ created under the Charter. Through this power of review, the Assembly can modify or eliminate any organ it wishes. In this way, it can amend the Charter informally without going through the long process of amendment. The AHG's power of review has been used in the past to eliminate the Commission of Mediation, Arbitration and Conciliation, which was replaced by more informal "executive" actions.

A second function relates to the Assembly's power over its own membership. The Charter empowers it to admit new members. The only qualification is that an applicant must be an "independent sovereign African state." And although it is not explicitly stated, it can safely be assumed that such an applicant state must generally agree with the principles of the Organisation. The Assembly through a simple majority may decide to admit or reject such applicant-state. However, the Assembly has never rejected the application for membership of any African state under indigenous control.[6] In the case of conflicting claims over representation of a state at the OAU, the AHG decides which of the claimants can represent that state, as it happened in Ghana (1966), Uganda (1971), and Chad (1983). Besides, it confers legitimacy on liberation movements as well as on mediation efforts of its members.

Finally, the AHG also performs elective functions. It is, for instance, responsible for the election of the Organisation's Administrative Secretary General and his five assistants. It also appointed the members of the defunct Commission of Mediation, Conciliation and Arbitration.

The Council of Ministers

The Council of Ministers (CM) is the implementing arm of the AHG. As such, it is directly responsible to the Assembly, functioning as its "cabinet." It is made up of the Foreign Ministers of member states or their representatives. It prepares the agenda of the conferences of the Assembly. In this way it can and does influence the agenda of the AHG; it debates and vets the policies recommended to the AHG. It serves to begin the process of deriving a consensus within the Organisation. However, its decisions are not binding on the Assembly. Indeed, the Assembly has in the past repudiated CM resolutions which it considered unacceptable. In 1965, the AHG rejected the CM resolution that called on all member states to sever ties with Britain if the latter failed to crush the Rhodesian rebellion by 15 December of that year.

Without any standing committee of its own, the Assembly relies on the CM to coordinate the general policies of the OAU member states in the political, economic, educational, health, scientific, and defence fields [Article 11 (2)]. This harmonisation and coordination is meant to be done through the directive of the AHG. However, there is hardly any effective coordination of the disparate and at times conflicting policies pursued by different OAU member states on African and other issues.

There is a statutory provision of biannual meetings of the Council. The first is generally held in February in Addis Ababa to discuss budget matters while the second is meant to prepare the agenda for the annual meeting of the Assembly. The Charter provides for an emergency meeting to be convened if two-thirds of all member states are in support. Emergency meetings were held in the past for a variety of reasons, ranging from border conflicts to colonial aggression. Resolutions of the Council of Ministers are passed by a simple majority of the members. However, highly controversial issues are usually passed on to the Assembly by the Council of Ministers for further and authoritative consideration by the heads of state.

The Secretariat

The OAU Secretariat is made up of international civil servants from the African continent. However, some of them serve at the pleasure of their governments, as in the UN system, so their purported neutrality is often negated.

At the head of the Secretariat is the Administrative Secretary-General, who is supposed to serve a four-year term which is renewable. Assisting him are five Assistant

Secretaries-General chosen to represent the geographical regions of the continent: East, West, North, South, and Central Africa. They are assigned to different functional areas, e.g., political, economic, administration, education, science, culture, and social affairs and finance.

The Administrative Secretary-General and his assistants are appointed by the Assembly. To win appointment, a candidate needs only secure two-thirds majority of all members present. There is no particular requirement for appointment. For instance, the candidate need not be bilingual nor does he or she have to be a Minister of Foreign Affairs. The prestige and status attached to this post has made it a hotly contested office, one which reflects the various ideological, regional and cultural factions within the OAU.

Since its creation in 1963, four Administrative Secretaries-General have served the OAU. Diallo Telli, a former Guinean Ambassador to the United Nations, served from 1964 to 1972. Although he sought re-appointment for a third term in 1972 he was defeated by Nzo Ekangaki, a Cameroonian nominated by Nigeria. The defeat of Telli could be accounted for by his expansive conception of the role of the Secretary-General and his militant views on colonialism. These ran contrary to the views of the many conservative and moderate states in the OAU.[7]

Nzo Ekangaki succeeded Diallo Telli as Secretary-General in June 1972. The good-will he enjoyed was shortlived, however; he resigned his appointment in 1974 following his controversial selection of Lonrho--a British multinational corporation with extensive interests in Rhodesia and apartheid South Africa--as consultants to the OAU on petroleum matters.

William Eteki Mboumoua was appointed to succeed his fellow Cameroonian in 1974. Mboumoua's selection was due to the deadlocked election between Omar Arteh of Somalia and Vernon Mwanga of Zambia. The Somali was backed by the Arabs and a few Francophone states; Vernon Mwanga, on the other hand, drew his support from the Anglophone states and Zaire, which resented growing Arab influence in the OAU. The resentment arose from what many African states perceived as excessive Arab influence in the Organisation. This is particularly so as Somalia had just then been admitted into the Arab League. But underlying this resentment was black Africa's bitter disappointment over the failure by the Arab states to sell oil to them at a concessionary price.[8] Mboumoua's term as OAU Secretary ended in 1978. He was succeeded by Edem Kodjo, another francophone who, before his election, was the Foreign Minister of Togo.

Such politics attendant on the appointments of OAU Secretary-Generals is just one of the problems faced by the OAU Secretariat. Other problems internal to the Secretariat include the issue of staffing. The OAU has not been able to attract or retain many competent and highly skilled staff. Member states who are themselves short of this type of men and women are reluctant to send their most qualified nationals to the Organisation. Another related problem is the fact that these appointments are rather "political." As a result, some staff at the Secretariat continue to maintain contacts with their home governments to the detriment of the Organisation. Moreover, it is impossible to discipline effectively errant Secretariat staff because of intervention from home governments. Third, there is a quota system which restricts the number of high-ranking officials any member state can send to the Secretariat. Of equal importance is the fact that OAU officials are poorly remunerated when compared with their UN counterparts, who are incidentally also based nearby in the Economic Commission for Africa's (ECA) headquarters in Addis Ababa, whose assembly hall is used by the OAU for its Assembly and Council meetings.

The OAU Secretariat is, among other things, responsible for the preparation of the Organisation's budget. The scale of assessment already noted is patterned after that of the UN, in which a country's income and its total population is used to compute the amount that should be paid by each member state. After preparing the budget, the Secretary-General submits it to the Council of Ministers for ratification. However, most member states do not honour their financial obligations to the Organisation for a variety of reasons ranging from disagreement with OAU policies to poverty of delinquent members. For instance, although the 1980 budget proposed by the Secretariat was a modest $17.6 million, over half (i.e., $9.1 million) of this was listed as outstanding from previous years. And unlike the UN neither the Secretariat nor the Assembly has the statutory power to punish defaulting states.

Commission of Mediation, Arbitration and Conciliation

The fourth major organ created by the OAU Charter was the now-moribund Commission of Mediation, Arbitration and Conciliation. The protocol of this Commission was approved by the OAU Assembly at its 1964 summit. Among other things, the protocol established a twenty-one-member commission to be headed by a President and two Vice-Presidents. All Commission members were to be legal practitioners in their respective countries, i.e. not normal "political" OAU participants. They were to be elected into the Commission by the AHG for a five year term which was renewable.

The modalities of conflict resolution were elaborately set out in the protocol of the Commission. Notwithstanding this fact, the Commission did not function as had been planned. There were many reasons for its failure to be effective, i.e., authoritative. The first relates to the fact that the consent of parties to a dispute is needed before the Commission could act. Lacking in mandatory jurisdiction, the Commission was unable to induce African states to use its rather comprehensive facilities. The second factor concerns the relative newness of African states. As new members of the international system, African nations are very jealous of their brittle sovereignty and political independence. Consequently, they perceive judicial arbitration as an erosion of their hard-won independence and sovereignty. A third consideration is that most African disputes are of a political nature—territorial irredentism, ideological and personality incompatibility, and so forth—which renders them problematic in terms of resolution.

Closely related to this last factor is the nature of one aspect of the Commission's work: arbitration, which is concerned mainly with points of law. Yet judicial actions are highly unpredictable. As such, African states are not prepared to submit their disputes before an international tribunal—continental or otherwise—unless they are absolutely sure of the validity of their legal positions. However, in most cases, this is impossible to ascertain beforehand. After all, territorial, irredentist, and personality conflicts are highly intricate and complex issues, not easily susceptible to clear-cut, let alone effective, legal interpretations. It is not surprising that, in the circumstances, African states have relied instead on other methods of conflict control where they could, as it were, put their cards on the table and still hope to improve their bargaining positions and to influence the final outcome of the peace process. The apparent ineffectiveness of the Commission, given the preference for extra-legal negotiation and mediation, led to its being down-graded to an ad hoc body in 1970. By the 1977 OAU summit, the Commission was completely eliminated and replaced with an OAU "Standing Committee on Disputes."

The African Liberation Committee

In order to speed up the realisation of the objective of liberating African territories still under colonial and racist minority regimes, the founding conference of the OAU set up a special committee, known as the African Liberation Committee (ALC). It was initially composed of nine members—Algeria, Egypt, Ethiopia, Guinea, Nigeria, Senegal, Tanzania, Uganda, and Zaire. However, the membership was subsequently increased to eleven and then seventeen.[9]

The main functions of the committee are (1) harmonisation of all assistance provided by African states for the liberation struggle and the management of the Special Fund that was set up for that purpose; (2) coordination of the efforts of the liberation movements; and (3) unification of liberation movements, where more than one exist, so as to enhance their effectiveness.

The OAU, as it was set up in 1963, represented the constrained victory of the moderates over the radicals in the Pan-African Movement. Consequently, its Charter reflected the urgent desire by the new African leaders to maintain the conservative political and territorial *status quo* which they had then just inherited from their erstwhile colonial masters. However, the broad consensus reached in 1963 has since given way to divisive interaction among the Member States. In addition the activities of some African states since 1963 have limited and in certain cases have even undermined the ability of the Organisation to fulfil its objectives. We shall in the following chapters examine in detail the OAU's performance since its inception in the following issue areas: decolonisation, strategic problems, economic development, and human rights.

NOTES

1. For detailed accounts of the Pan-African Movement, see Immanuel Geiss, The Pan-African Movement (London: Methuen, 1974); Edmund David Cronon, Black Moses (Madison: University of Wisconsin Press, 1964); Colin Legum, Pan-Africanism: A Short Political Guide (London: Pall Mall, 1965); Adekunle Ajala, Pan-Africanism: Evolution and Progress (London: Andre Deutsch, 1973); Vincent Thompson, Africa and Unity (London: Longman, 1969); and Immanuel Wallerstein, Africa: The Politics of Unity (New York: Vintage Books, 1969).

2. For more details on intra-bloc politics, see Thompson, Africa and Unity, and Legum, Pan-Africanism.

3. Keesing's Contemporary Archives (KCA), 15-22 June 1963, p. 19463.

4. For the number of emergency sessions held since 1963, see Appendix 3.

5. At the Nairobi OAU summit in June 1981, there was opposition from Egypt, Senegal, Sudan and some other African countries, to the "choice" of Libya as the next venue for the 1982 summit. However, their opposition did not prevent the OAU from adopting Libya as venue for its 19th annual summit.

6. The OAU has in the past rejected the credentials of particular regimes, though. For instance, Tshombe in

1964, and most recently, the regimes of Master Sergeant Samuel Doe of Liberia (April 1981) and Hassene Habre of Chad at the Tripoli "non summit" in 1982. All regimes were subsequently accepted by the Organisation.

7. For more details, see Michael Wolfers, Politics in the Organisation of African Unity (London: Methuen, 1976), p. 64; and Zdenek Cervenka, The Unfinished Quest for Unity: Africa and the OAU (New York: Africana, 1977), pp. 31-32.

8. Olusola Ojo, Africa and the Arab World (London: Rex Collings, 1982), Chapter 6.

9. The current members of the ALC are Algeria, Cameroon, Congo, Egypt, Ethiopia, Ghana, Guinea, Libya, Mauritania, Morocco, Senegal, Somalia, Tanzania, Uganda, Zaire, and Zambia.

2
Decolonisation

Many African states attained formal independence at
a time when the world system encouraged the idea of
popular sovereignty and the independence of peoples.
The modality of independence that stressed negotiation,
harmony and continuity of economic forms influenced very
greatly the initial tactics and strategies for the
liberation of the remaining colonial territories in Africa.
This chapter is an analysis of OAU support of decolonization in Africa with great emphasis on the strategies
pursued and the implementation of those strategies in
Rhodesia, the Portuguese territories, the "trust
territory" of Namibia (South West Africa), and South
Africa. Although South Africa is legally independent,
it is nonetheless ruled by a white minority regime and so
can be considered the most important territory, given its
support for white minority rule in the other territories.
Africa's concern with colonial and white minority
rule in the continent stems from a variety of sources.
First, as noted already in Chapter 1, the new leaders of
independent Africa believe that their own independence,
freedom, and security are indivisible from those of their
brothers still under colonial rule. Hence they did and
do believe that their own freedom and security are in-
complete so far as any section of the continent is under
foreign and/or racist domination. This is understandable
given the objectives of the Pan-African movement of which
the OAU is the institutional consummation. And, second,
the presence of colonial and white minority and racist
regimes in the continent is seen as an insult to the
integrity of the black race. The founding conference of
the OAU, as might be expected, declared that colonialism
"is a flagrant violation of the inalienable rights of the
legitimate inhabitants of the territories concerned" and
"a menace to the peace of the continent."[1] The assembled
African leaders accordingly called upon the colonial
powers to grant immediate independence to their
territories.

The environment in which the OAU decolonisation policy was made influenced very greatly the strategies pursued by the Organisation. The pervasive characteristics of dependence, underdevelopment and military weakness have served to limit the effective options available to the Organisation in its quest for decolonisation. Subsequently the OAU has identified, adopted, and refined the strategies of moral suasion; the application of sanctions--diplomatic, social, and economic; and armed struggle. We shall discuss these strategies in some detail in the next section, given the constrained environment of the OAU.

MORAL SUASION

Most African states that had by 1963 attained their independence, especially the former British colonies, had done so through a mixture of persuasion, negotiation, and gradual devolution of political power from the colonial metropole. Leaders of independence movements were thus convinced that such tactics could serve as a model for the rest of the continent. They believed that moral pressure applied through the UN--in which they had great faith as well as large numbers--and through talks would accelerate the process of decolonisation. Besides such political preferences, many of those leaders whose views formed the bedrock of the OAU were conservative men who eschewed militancy and who believed that colonialism was beneficial to Africa. For example, the prime minister of Nigeria, Tafawa Balewa, who was then quite influential in the OAU, declared at his country's independence that

> We are grateful to the British . . . whom
> we have known first as masters, and then as
> leaders, and finally as partners, but always
> as friends. . . . We are grateful to those who
> have brought modern methods of banking and of
> commerce, and new industries.[2]

Similarly, President William Tubman of Liberia, who as ex-leader of the Monrovia faction also had tremendous influence within the OAU at that period, believed that Britain "has made tangible long-term economic contributions towards the political security and future prosperity of those areas over which she long exerted paternalistic authority."[3]

At the time the OAU was founded, not only were the constituent African states militarily weak but they also lacked the institutional and economic base to contemplate or support military confrontation with either the colonial powers or the regimes of the white redoubt. There was, and still is, no unified military command structure; the moderate African states had early and flatly rejected Nkrumah's suggestion of a Pan-African military command that could be used to liberate dependent territories.

This cluster of reasons influenced the adoption of moral suasion as a central strategy of liberation by the OAU. By and large the conservative and moderate majority in the Organisation considered that if Britain, a great power, could be persuaded to give independence to its colonies, then surely Portugal, a weak and poor state, would need less persuasion. This reasoning was implied in the position paper presented by Nigeria to the OAU in May 1963. Significantly, the OAU Liberation Committee endorsed the Nigerian proposal, which urged that Portugal be persuaded to accept the principle of self-determination for its African territories.[4]

Nevertheless, as early as 1960 individual territories, particularly those in the "radical" camp, had already embarked upon strategies other than moral suasion. For example, there were already on-going liberation wars in Algeria and the Portuguese territories. Besides, interstate organisations which preceded the OAU, such as the All African Peoples Organisation, had in the late fifties and early sixties called for the immediate application of political and economic sanctions collectively and individually against the government of the Union of South Africa.[5]

SANCTIONS

International organisations have always considered and sometimes used sanctions in their efforts to attain desirable collective objectives. Both the League of Nations and its successor organisation--the United Nations--have been prepared to apply sanctions against recalcitrant members who contravened universally accepted norms of state behavior. The OAU is not an exception; it has equally tried to use sanctions to bring about desirable changes in colonial and racist territories.

Socio-political and economic sanctions refer to the threat and threshold of non-engagement or the severance of social and economic intercourse with the regimes of the white redoubt. The imposition of such sanctions is intended to punish the regimes for not sharing universally accepted norms of self-determination for colonial peoples and the equality of races. African leaders hoped, initially, that economic and diplomatic isolation of colonial and racist regimes in the continent would force these regimes to decolonise and/or grant equality to all races.

Socio-political isolation and moral suasion are of course "soft" strategies of liberation. Although economic sanctions could be categorised as a "hard" strategy if applied by a dominant actor against a dependent state, i.e., France versus one of its "dependents"--Central African Republic--this was not the case with Africa's economic sanctions against colonial and racist regimes simply because of the dependent nature of the African states and the limited economic links between them and the target regimes. This latter condition is not, of course, so for those Front Line States neighbouring on South Africa, whose attitude towards sanctions is very much affected by their multiple dependencies on the white "semi-periphery" of South Africa.

ARMED STRUGGLE

Armed struggle was not initially emphasized as a liberation strategy because of the naive faith of African states in peaceful decolonization. The only reference to armed struggle in 1963 was indirect. The OAU did call on all members to provide "material assistance" to the liberation movements; it was clear, however, given the low level of members' mobilised military resources that the Organisation had no intention of engaging in a direct test of arms with the colonial and racist regimes. Rather, the OAU was to rely on nationalist freedom fighters in the various territories to bear the brunt of the military confrontation, i.e., guerrilla struggle. They, however, established the African Liberation Committee to coordinate assistance to the liberation movements.[6]

IMPLEMENTATION OF SANCTIONS

At its inaugural meeting the OAU adopted two resolutions asking for the severance "of diplomatic and consular ties between all its members and the government of Portugal" and the imposition of "an effective boycott of the foreign trade of Portugal." The focus on Portugal is predictable as it was thought to be the weakest link in the colonial chain. It was also assumed that the independence of Portuguese territories would have a domino effect on other colonial territories. Two questions arise, however, as to (1) how many African states actually heeded the call of the organisation and (2) the effectiveness of the sanctions; that is, whether the target states were affected by OAU goals.

When the OAU was founded in 1963, very few African states had diplomatic and social relations with Portugal. Several states in Africa may have been constrained at independence by resolutions of the United Nations and earlier Pan-African bodies which called for the ostracism of this regime and its colonies. These two reasons taken together may have been responsible for the massive support by many OAU states for the diplomatic and social isolation of Portugal; given limited historical contact, such "sanctions" were easy to accept and effect. Except for Malawi, which opened an embassy in Lisbon in November 1969, and Morocco, which maintained diplomatic relations with Portugal until 1972, all other African states had by 1970 complied with the OAU demand for the diplomatic isolation of the Caetano regime in Portugal.

The proposed economic boycott of Portugal, on the other hand, was less successful. Some member-states of the OAU did not faithfully observe such sanctions as imposed by the Organisation. Prominent among the economic sanctions violators were Gambia, Liberia, Malawi, Morocco and Nigeria. In September 1973, the Movement for Justice in Africa (MOJA) in Liberia revealed that the official residence of the country's President was importing Portuguese wines for its kitchen.[7] Between 1963 and 1970, Nigeria exported U.S.$72.3 million worth of goods to Portugal while importing U.S.$8.6 million from it,[8] so Nigeria had a bilateral trade surplus of U.S.$64.24 million in its favour. The benefits accruing to African states from continued economic links with Portugal was partly responsible for their breaking OAU sanctions. It is also quite possible that some other OAU states violated the economic boycott of Lisbon without being known. And some might have continued their economic links with Lisbon through third parties.

Concurrent with the implementation of economic sanctions, no matter how complete or effective, was the pursuit of the armed struggle in the Portuguese held territories. However, in the absence of a Pan-African force, the OAU could not play a direct and effective role in the armed struggle. This was partly by design and partly by default, the primary responsibility of the nationalist movements in the various territories. Nevertheless, the OAU did play a role through the ALC.

The ALC provided its assistance to the nationalist movements in the lusophone and other territories in four different ways. First, it extended recognition to the liberation movements. Such recognition confers three advantages on the movements: (1) it legitimises their struggle; (2) it gives international acceptance; and (3) it usually results in increased aid. However, such recognition is always conditional: it can be withdrawn if the movement is perceived to be fractionalised and ineffective in prosecuting the liberation struggle. In 1963 the ALC conferred exclusive recognition on the National Liberation Front of Angola (FNLA) under the patently pro-American Holden Roberto, who had established the Revolutionary Government of Angola in Exile (GRAE).[9]

However, the exclusive recognition was withdrawn in 1964 because of the existence, by then, of the more effective Popular Movement for the Liberation of Angola (MPLA).

A second function of the ALC is its mediatory role between rival liberation movements. For instance, several unsuccessful attempts were made to reconcile Roberto's GRAE and the MPLA, led by the late Agustinho Neto.

Third, the ALC, as its name suggests, attempts to co-ordinate assistance to liberation movements. For instance it disburses the OAU Liberation Fund to the nationalist movements. However, not all funds go through the ALC. This is because the bulk of assistance to the liberation movements is provided by sources outside Africa which prefer to deal directly with the movements. Similarly, even certain African states, despite their OAU or Liberation Committee membership, prefer bilateral relations with movements which they deem effective or with which they have a particular ideological sympathy. Furthermore, donors prefer direct access to the liberation movements and anticipate favours after the liberation struggle had been won.

Despite its variety and controversy, OAU assistance did not make any decisive impact on the armed struggle. Indeed, besides diplomatic support, which is rather easy to give, OAU assistance to the liberation movements was rather feeble, inadequate, and intermittent. A con-sequence of this was the elusiveness of effective insti-tutional authority by the OAU over the movements.

Many reasons can be adduced for the Organisation's poor performance in promoting the liberation struggle. First, is the lack of real commitment to this struggle by many OAU member states. Commitment varies according to the ideological disposition of the African states and the distance from the theatre of conflict. "Radical transformationist" states such as Algeria, Ghana (under Nkrumah), and Tanzania are more dedicated to the liberation struggle. Conversely, conservative peripheral capitalist states like Gabon, Ivory Coast, Senegal, Tunisia, and Zaire are far less committed and constitute retrograde critics of the ALC.

Second, and closely related to the first point, is the inadequate financing of the ALC. For example, between 1 December 1971 and 31 October 1972 contributions received from member states towards the Special Fund were only £548,058.[10] Most African states are always in arrears of their contributions to the Liberation Fund and the OAU budget. In 1970, only twelve members had paid their assessed contributions to the Fund; in 1975 this number had been drastically reduced to five,[11] and by the end of November 1981, only three countries--Guinea, Nigeria, and Zambia--were not in arrears. Even states which could have single-handedly funded the Liberation Committee, given sufficient political will, were them-selves delinquent. Egypt, for instance, owed over

$2.4 million in November 1981.[12] Moreover, the total
budget of the ALG is not only small, but a substantial
part of it is spent on administration rather than on
assistance. In the period between 1 December 1971 and
31 October 1972, for instance, the ALC expended
£470,448.16. Out of this amount only £151,349.21, or 17
percent of the amount earmarked in the budgetary appro-
priations, went directly to the liberation movements.[13]
The rest was spent on the maintenance of the training
centres in East and West Africa, office annex extension
and furnishing, the OAU Trade Fair, construction of ware-
houses, the coordinating office established for the training centres in East Africa, and the personal emoluments
of the military experts.

Finally, the ALC itself was riddled with a myriad of
problems. These range from allegations of maladministra-
tion and conflicts between the "radical" and "conservative"
elements within the Committee to its large membership,
which makes it an unwieldy, inefficient, and cumbersome
decision-making body.

DECOLONISATION OF LUSOPHONE TERRITORIES

Guinea-Bissau

The failure of the OAU in the decolonisation "issue-
area" was dramatically brought home to the Africans by
the processes through which the Portuguese territories
actually attained their independence.[14] Guinea-Bissau,
for instance, unilaterally declared itself independent
of Portugal in September 1973 after a long and bitter
guerrilla war in which the OAU offered minimal material
assistance. As a matter of fact, the African Party for
the Independence of Guinea and Cape Verde (PAIGC), the
major nationalist movement in Guinea-Bissau, had
complained about and condemned the paltry assistance
which the OAU had afforded it. Instead of effective OAU
challenges, it was the human and material cost of the
protracted struggle for Portugal, a resource-poor
colonial metropole, which led to internal political
turmoil in the country which culminated in the April 1974
coup against the regime of Caetano.

Before its overthrow, the Caetano regime, which was
losing the colonial war in Guinea-Bissau, more conspicuously made a desperate attempt to oust President
Sekou Toure of Guinea, then a foremost anti-colonial
leader and an ardent supporter and ideological ally of
Amilcar Cabral's PAIGC. On 22 November 1970 Portuguese
soldiers aided by dissident Guineans invaded Conakry.
Their aim was not only to overthrow Sekou Toure but also
to capture Amilcar Cabral, who was then living in the
Guinean capital. The OAU reaction to Portugal's violation
of its Charter and that of the UN was predictably

inadequate. It consisted essentially of anticipated and
routine resolutions by the Council of Ministers condemning Portuguese aggression against Guinea; there was no
material or financial assistance to either Guinea or the
PAIGC. However, the OAU in a feeble symbolic gesture
decided to establish a regional office of the Liberation
Committee in the Republic of Guinea. This office was
opened in November 1973. Yet, in December of the same
year, the OAU accepted into membership the new Republic
of Guinea-Bissau. By this admission, the OAU conferred
legitimacy on the new state and so made its new office
superfluous.

Angola

Unlike other Portuguese territories in Africa, no
nationalist movement was dominant in Angola. The OAU's
main role there was to seek to effect reconciliation
between ideologically incompatible and ethnically
disparate factions. However, its lack of resources to
assert its authority was partly responsible for the
Organisation's failure to unify the nationalist movements.
This failure was patently demonstrated in 1975 following
the Portuguese decision to grant independence to the
territory. Various unilateral and multilateral initiatives were made by African Heads of State as well as by
OAU officials to secure a government of national unity.
The late Jomo Kenyatta "put his prestige on the line"
when he summoned a meeting of the leaders of the three
warring factions, namely Agustinho Neto of MPLA, Holden
Roberto of FNLA, and Jonas Savimbi of the National Union
for the Total Independence of Angola (UNITA). The three
nationalist leaders re-endorsed the commitments to the
Alvor Agreement signed with Portugal. Under this Agreement, Neto, Roberto, and Savimbi agreed to work with
Lisbon in a coalition government during the transition to
independence. The failure of the parties to adhere to
this agreement led Kenyatta to call a second meeting of
the leaders at Nakuru in June. Although they agreed to
restore a climate of peace in Angola, the renewed
understanding collapsed immediately for a variety of
reasons.[15]
First, while the MPLA is a socialist party with close
links to the Soviet Union, the FNLA's patron is the
United States of America, which channeled its support
through its more reliable client, Zairean leader Mobutu.
Ideological differences apart, the movements were divided
along ethno-regional lines. The FNLA draws its support
from the Bakongos in the North; the MPLA is backed in the
main by the Mestizos and others (not a "tribe" as such)
in the urban areas; and finally, the UNITA commands the
allegiance of the Ovimbundus in the South.

Second was the prize at stake. The leaders were cognisant of the fact that politics in Africa tends to be a "zero-sum game" in which the winner scoops everything while the loser is sent to jail or forced into exile, if not assassinated. From that perspective the Angolan leaders did not want to lose given the penalty for failure, yet they could not bring themselves to cooperate to share the benefits of their struggle.

While Kenyatta's initiatives were made in his personal capacity as an elderly African statesman, the OAU made collective institutional efforts to supplement and advance Kenyatta's peace moves. At its July 1975 Assembly, the Organisation established a conciliation commission to reconcile the warring factions. The Commission met on 5 November and proposed a government of national unity embracing the three liberation movements. This peace package was however rejected by the MPLA for two main reasons: (1) it was militarily the strongest of the three liberation movements and was therefore reluctant to share power with the others; and (2) it was already in control of the capital, Luanda, and therefore strategically placed to inherit the apparatus, if not all the territory of the state about to be abandoned by the Portuguese. The OAU peace plan was thus perceived as another stumbling block in the attainment of this objective.

On 11 November 1975 Portugal withdrew from Angola as promised. This "scuttle" created a vacuum which the OAU could not fill through the forging of a national unity government of the three movements. Indeed, member states of the Organisation were divided along "conservative" and radical "lines" over their preference for different nationalist movements. The task of the OAU was not made any easier by the establishment of rival governments, not just parties, by the nationalist forces: the MPLA in Luanda and the FNLA-UNITA coalition in Umbarto. Each rival government sought to gain support from its patron states as well as collective legitimisation from the OAU. The MPLA, for instance, solicited and obtained assistance from Cuba and the Soviet Union. The FNLA-UNITA coalition, on the other hand, obtained covert support from the USA as well as open assistance from China, Zaire and the state which most African states consider as their foremost enemy, South Africa.

Increasing external intervention—albeit at the "invitation" of alternate "regimes"—not only exposed the Organisation's inability to bring the rival nationalist leaders together; it also forced the OAU to make another effort aimed at settling the issue of recognition for the rival governments now established in Angola. In January 1976, the first extraordinary meeting of the AHG was held to discuss the Angolan problem. But by the end of this two-day summit, member states could not find a solution to the conflict. There were two conflicting positions. While the "conservatives" led by Senegal insisted on a government of national unity, Nigeria, in one of its rare radical stances, joined the "radicals" to demand immediate recognition for the MPLA. Nigeria under the post-Gowon military regime, also requested the Organisation to denounce the armed aggression of "racist and fascist" regimes in collusion with FNLA and UNITA mercenary contingents."[16]

But the mutually exclusive positions of the two groups of states produced a predictable deadlock. A few weeks after the summit, the OAU put its seal of approval on the MPLA government when it admitted Angola into the Organisation. Nigeria, by "switching" from conservative to radical "camp" had started a bandwagon effect resulting in MPLA diplomatic "victory."

RHODESIA

The Rhodesian problem was slightly different from that of the Portuguese territories. While the Portuguese territories were treated as cases of political decolonisation, the Rhodesian situation was seen as both a decolonisation issue as well as the abolition of institutionalised racism. Since Britain was held to be responsible still for the territory, the OAU decolonisation strategy of moral suasion was accordingly directed towards that metropolitan country. This continued until the Unilateral Declaration of Independence (UDI) in November 1965. Before then, at the founding conference of May 1963, member states had appealed to Britain "not to transfer the powers and attributes of sovereignty to foreign minority governments imposed on African peoples by the use of force and under cover of racial legislation."[17]

The tension between white minority rule and the demands for black majority rule in the constituent regions of the Central African Federation, which was formed in 1953, led to the collapse of that Federation in late 1963. The white settlers in Southern Rhodesia, cognisant of the independence agitation of the other parts of the Federation, braced themselves for UDI. The Rhodesian minority government severely repressed the nationalists in order to forestall anticipated mass African opposition to this impending illegal action. This brutal repression as well as the anticipated UDI prompted the OAU Council of Ministers in February 1964 to call once again on Britain to prevent the "threat of unilateral independence or subtle assumption of power by the minority settler regime in Southern Rhodesia."[18]

By mid-1964 the OAU effort to convince Britain to prevent UDI had clearly failed. The recognition of this

fact led the Organisation to harden its attitude towards
Britain. The Organisation's hardline was its recognition
of the failure of moral suasion. Consequently, resolutions
started to raise the prospect of sanctions against Britain
and the territory. In August 1965, the AHG called on
member states "to reconsider all political, economic,
diplomatic and financial relations with the Government of
the United Kingdom of Great Britain and Northern Ireland
in the event of this Government granting and tolerating
the independence of Rhodesia under a minority Government."
The Organisation threatened the use of force against
the minority regime (but not against the metropole) in the
event of UDI.

However, when Rhodesian Premier Ian Smith unilaterally declared the colony independent in November 1965
the OAU was incapable of using force to dislodge Ian Smith
and his illegal administration. This was because the
OAU had no standing army nor could it assemble an <u>ad hoc</u>
military force to combat Ian Smith. Rather than use force,
the Council of Ministers met in an emergency session in
Addis Ababa in early December and resolved "that if the
U.K. does not crush the rebellion and restore law and
order, and thereby prepare the way for majority rule in
Southern Rhodesia by December 15, 1965, the member states
of the OAU shall sever diplomatic relations on that date
with the United Kingdom."[19] But when the ultimatum
expired on the 15th, only nine out of the thirty-six
states in the OAU then had complied with the resolution.
These were mainly the militant anticolonial states who
were ready to forego the crumbs of economic assistance
from Britain for the sake of African freedom. The majority
of OAU states under conservative and pro-Western regimes
did not implement the resolution partly because they were
reluctant to antagonise Britain and partly because they
had only a feeble commitment to decolonisation. Anglophone states in the Commonwealth had another reason for
not severing ties. Most of them were recipients of
British aid, which would have been adversely affected by
any diplomatic rupture.

The reluctance of the British government to use force
against a regime with which it had racial affinity, and
the inability of the OAU to implement its own resolutions
on Rhodesia, led the Organisation to take the issue to
the United Nations, i.e., to escalate the issue from a
regional to a global level. Before UDI African states
had got the United Nations involved in the matter of
Rhodesia. The African group had successfully convinced
the General Assembly that contrary to Britain's assertion,
Rhodesia was not a self-governing territory.[20] However,
the African bloc was not able to convince the Security
Council to impose mandatory sanctions until May 1968,
when it prohibited trade with, investment in, and travel
to Rhodesia.[21]

The political support of the UN for the OAU on the Rhodesian problem did not bring about the collapse of UDI, partly because the Africans could not convince the Security Council to extend sanctions to those states whose compliance was needed to make them successful--the neighbouring countries of South Africa and Portugal rather than the metropole of Britain.

Whilst the OAU was pressing for sanctions against the known and open sympathisers of Rhodesia, Britain held intermittent talks with the illegal regime with the intention of bringing it to legality. The talks, held variously in 1966, 1968 and 1971, did not resolve the problem, however. The OAU position on these talks was that there could not be an internationally acceptable solution to the Rhodesian problem which did not grant majority rule to Africans before independence.

From 1974 onwards, the OAU passed on the responsibility of bringing Rhodesia to legal independence to the Front Line States.[22] The strategic location of these states, plus their shared ideological beliefs with various factions in the nationalist movements, made them influential in the liberation struggle in Southern Africa. Besides these, some of the Front Line States had hitherto played a prominent role in the formulation of OAU liberation strategies. Tanzania, for instance, not only houses the ALC headquarters, it also provides the Executive Secretary of the Committee.

A dramatic increase in the influence of these Front Line States occurred after the coup in Portugal in April 1974. The subsequent independence of Mozambique and Angola also had a traumatic impact on the minority regimes in Southern Africa, particularly Rhodesia. Strategically, it exposed the Smith regime to more direct military attacks by the nationalists. This was so because the coup added some 500 kilometers to Rhodesia's already extensive defence perimeters. On the economic front, the liberation of Mozambique opened Rhodesia to more effective economic sanctions which had not been possible before because Portugal had collaborated with Ian Smith to bust sanctions.

Essentially, the Front Line States, as de facto "OAU agents," embarked upon two broad policies within the framework of the Organisation's established strategies. First, they sought to unify as well as to legitimise movements that were perceived to be effective fighting forces and which were similarly interested in genuine black majority rule. And second, they set out the modalities for the resolution of the regional conflict.

First then, Front Line States as unifiers and legitimisers brought about the integration of the factions in Zimbabwe into the African National Council under Bishop Abel Muzorewa in December 1974. Yet within three years the continuing centrifugal forces in the ANC led to

its collapse. The resultant factions of Robert Mugabe's
Zimbabwe African National Union (ZANU) and Joshua Nkomo's
Zimbabwe African Peoples Union (ZAPU) formed the
Patriotic Front (PF). The Front Line States, among other
reasons, endorsed this new alliance because of its
militancy towards the Smith regime.

The second major function of the Front Line States
as noted was concerned with setting out acceptable
modalities for resolving the Rhodesian problem. In the
main, these were no independence before majority rule and
no acceptable settlement without the fullest participation
and consent of the Patriotic Front and its fighting
forces. Given these preconditions, then, the Front Line
States rejected the Kissinger plan. The Kissinger plan,
among other things, would bring about "majority" rule in
two years. In addition, the plan called for a transi-
tional administration with an equal number of whites and
blacks in the Council of State, whilst the blacks would
form the majority in the Council of Ministers. However,
the strategic ministries of Defence and Law and Order
would be held by whites.[23] This would have compromised
black majority rule. Consequently, the Geneva talks
which followed the American plan, as might be expected,
collapsed.

Following the collapse of these Geneva talks, the
initiative for resolving the Rhodesian problem passed on
to the Commonwealth. The most important breakthrough
came at the 1979 Lusaka meeting of the distinctive international Organisation. Through pressure from African
members, particularly Nigeria, Tanzania, and Zambia, the
new British Prime Minister, Margaret Thatcher, who
apparently had been predisposed to the internal settlement of Smith and Muzorewa, modified her stance. This was
reflected in the conference's decision that constitutional
talks under the chairmanship of Britain be held, intended
to bring the territory to legal and acceptable independence.

When these constitutional conferences opened at
Lancaster House on 10 September 1979, the OAU was not a
major participant. Its role was limited to that of an
observer. The agreements reached in London among other
things provided for general elections under Commonwealth
observation. The elections that were held under the
Lancaster House provisions brought to power the ZANU-PF
faction of the Patriotic Front led by Robert Mugabe in
April 1980. The new state of Zimbabwe became the fiftieth
member of the OAU soon after its independence.

NAMIBIA

At first sight, Namibia, South West Africa, appears
to be a straightforward case of decolonisation. It is a

trust territory whose independence was practically assured according to Article 76 of the UN Charter. However, the complicating factor has been the refusal of South Africa to hand over its League mandate to the United Nations Trusteeship Council,[24] according to Article 77 of the Charter.

The collective African interest in Namibia became prominent after the first wave of independence in the early 1960s. In 1960 Liberia and Ethiopia, the only African members of the defunct League of Nations, sought on behalf of the continent the opinion of the International Court of Justice (ICJ) on the legality of South Africa's continuing presence in South West Africa. Whilst the proceedings dragged on the OAU was created. The Organisation quickly put its stamp of approval on the litigation commenced by the two African states. But contrary to international expectations, the Court ruled that the two African states "were not entitled to a judgement on the validity of their claim."

The OAU's reaction to the Court's unhelpful decision was rather predictable. It reacted by endorsing the UN General Assembly's decision terminating South Africa's mandate in the territory. However, the OAU accepted a subordinate role over the issue by pledging "wholehearted co-operation with the United Nations in discharging its responsibilities with respect to South West Africa."[25] This secondary position was to be expected. First, the UN had taken direct responsibility for the territory in its October 1966 resolution. Second, and more importantly, the OAU lacked resources to play a direct role in Namibia, given in particular its concentration on Angola and Mozambique at the time. OAU activity on Namibia has by and large consisted of (1) symbolic material assistance to SWAPO Freedom Fighters; (2) passage of resolutions calling for the independence of the territory and condemning South Africa's reluctance to hand it over to the UN; and (3) recognition and legitimisation of SWAPO as the representative of the people of Namibia.

The stalemate produced by South Africa's refusal to abide by the UN resolutions demanding its withdrawal from Namibia and the OAU's impotence in expelling South Africa from the territory, led to the emergence of the "Contact Group" of five Western countries--United States, West Germany, Great Britain, France, and Canada--acting on behalf of the UN Security Council. The aim of the Contact Group is to bring about a peaceful transition to acceptable legal independence so ensuring continued Western influence and access to the territory's rich mineral resources.

But progress towards any peaceful settlement has been painfully slow. There have been a variety of reasons for this. First, there are the fears in the West about the

possible domination of an independent Namibia by the
socialist-inclined SWAPO. And, second, is the emergence
of a conservative American administration under Ronald
Reagan which perceives South Africa to be a strategic
link in the Western defence chain. Given these reasons,
the West under the influence of the United States has been
reluctant to apply necessary pressure to force South
Africa to implement speedily the UN time table for the
independence of Namibia. Meanwhile, the OAU has been
compelled to watch these unfavourable developments which
it has no resources to influence decisively other than
issuing statements condemning those actions by South
Africa which it considers to be contrary to the interest
of Africans.

SOUTH AFRICA: THE LIMITS OF ORTHODOX DECOLONISATION

South Africa's wealth, its level of industrialisation, and its strategic location make it the highest prize
that could be plucked by Africans in their continuing
quest for total decolonisation, elimination of racism,
and rapid economic development of the continent. With a
Gross Domestic Product of $53 billion in 1981, South
Africa is the richest country on the continent. However,
this wealth does not circulate evenly among the country's
population. Given the fact that the riches of the
Republic are shared by a handful of 5 million whites out
of a population of 29 million people, mainly blacks, it
is not surprising why there is strong opposition against
the South African racist regime among African states.
South Africa, however, is not a classic case of
decolonisation. The major goal of the African states is
not the independence of the Republic per se. Rather, it
is to effect a change in regime--the transfer of
political power from the white minority to the African
majority--and to put an end to racial discrimination
concretised in the officially sanctioned policy of
"separate development" or apartheid.
Global organisations are committed to the equality of
the races and do abhor racial discrimination. The UN in
particular proclaims the equality of all races in its
Declaration on Human Rights. Member states of the UN,
India in particular, had on several occasions drawn the
attention of the world body to the discriminatory policies
of the South African government against blacks, including
Coloureds and Asians. In the 1960s, the UN Security
Council adopted a resolution which considered South
Africa's racial policy to be a danger to international
peace and security. By 1963, then, an anti-apartheid
tradition had clearly been established in the UN and many
other international forums.

As earlier stated, the OAU adopted various strategies for the elimination of minority rule and apartheid on the continent. As in the case of Portugal and Rhodesia, many member states implemented the OAU call for the diplomatic and economic isolation of South Africa. All African states except Malawi have refused to exchange diplomatic relations with the Republic, and have also withheld recognition of the pseudo-independent "Bantustans" created by the apartheid regime. Similarly, most African states have refrained from conducting open trade relations with Pretoria. The significant exceptions, apart from the colonially-influenced trade links of Botswana, Lesotho, Mozambique, Rhodesia, Swaziland, and Zambia are Central African Republic, Gabon, Ivory Coast, and Zaire. In 1979, for instance, Zaire imported goods worth $131 million from South Africa, which represented a quarter of South Africa's total exports to the rest of Africa.[26]

African states carried their campaign to isolate South Africa to international forums other than the OAU and the UN. For instance, in 1964, they succeeded in expelling South Africa from the International Labour Organisation. South Africa was equally pressured out of the International Atomic Agency in 1976. The most far-reaching attempt to isolate South Africa occurred in October 1974 when its delegation was barred from participating in the UN General Assembly session of that year. Indeed, South Africa would have lost her UN seat had it not been for the "protective veto" cast by Britain, France, and the United States in the Security Council.

Apart from hounding South Africa from international forums, the OAU imposed a sports boycott on the apartheid regime. For instance, the Organisation has prohibited direct sporting and cultural links between Pretoria and African states.[27] In addition, the OAU has also imposed a "secondary" sports boycott on other non-African countries which maintain sporting ties with the Republic. This secondary boycott became necessary for two important reasons. First, the possibilities of direct boycott had been exhausted: every African state complied with the OAU sports sanctions call. And second, primary boycott had only been partial in that it did not involve other extra-continental states which maintained sporting ties with South Africa. By imposing a secondary boycott, the African states were seeking to impose a global sporting ban on South Africa whose nationals, particularly those in the white community, are sports enthusiasts.

Although many African states implemented the diplomatic, social, and economic sanctions imposed by the OAU, this compliance did not and could not induce significant and fundamental changes in South Africa. There are many reasons for this. First, as indicated, South Africa is not an orthodox case of political decolonisation in which political power passes to a black elite which maintains

the economic substructure intact. Second, the ruling
Afrikaners regarded themselves as Africans--a "white
tribe"--whose stay in Africa is not only legitimate but is
also expected to be continuous, for all time. In the
perception of these Afrikaners, the transfer of power to
black Africans would not only mean the loss of political
power but would also endanger their cultural, religious,
and social identities. Finally, the support of Western
capitalist states which have extensive investments in,
strategic military interest in, and racio-cultural ties
with South Africa has thwarted OAU and UN attempts to
effect meaningful change in South Africa.

While social and economic sanctions were being proposed and partially imposed on South Africa, the OAU also
pursued the armed struggle option through the liberation
movements. It has given moral and material assistance to
the main liberation movements: the African National
Congress (ANC) and the Pan-African Congress (PAC). This
support, however, has been minimal. For instance in 1973,
the ANC and PAC were jointly allocated 5 percent of ALC
budget (about U.S.$175,000). This is indeed a paltry sum
compared with South Africa's defence budget of U.S.$448
million for the same year.[28]

This minimal financial commitment of the OAU to the
liberation movements--a reflection of Africa's poverty--
coupled with the apparently overwhelming military strength
of South Africa, led some member states of the Organisation to question the efficacy of its liberation strategies
with regard to the Republic to advocate dialogue and
detente rather than antagonism.

President Houphouet-Boigny of Ivory Coast suggested
in 1970 that the OAU should abandon the confrontation
strategy, which he said had woefully failed to bring
about change in South Africa. The Ivorian leader's
arguments were ostensibly based on the Lusaka Manifesto,
which advocated a dual strategy of "talk and fight."[29]
Houphouet-Boigny's preference for talk with South Africa--
which he had been doing secretly--was compatible with
his political ideology. A big plantation owner and
medical doctor, Houphouet-Boigny is passionately anti-
communist and unabashedly pro-West. He was thus averse
to what he saw as the domination of the ALC and the OAU
liberation strategy over South Africa by the radical
African states.

However, neither Houphouet-Boigny nor his conservative supporters in the OAU specified what the Organisation had to offer in any negotiations with South Africa.
Indeed, "dialogue"--in the form of moral suasion--had
been tried unsuccessfully in the past. Given this, it is
doubtful whether the "new dialogue" would have achieved
the desired objective of majority rule in South Africa.
This point is important in the light of South Africa's
precondition that its internal political arrangements were

not to be subject to discussion. This was perhaps the
reason why the OAU in its 1971 summit firmly rejected
dialogue with South Africa. The majority of member states
reasoned that dialogue with South Africa would be a
negation of the principles of the OAU Charter and would
imply a rejection of the Lusaka Manifesto.[30] Besides,
acceptance of dialogue without radical concessions from
South Africa, such as a readiness to consider or to
concede to the complete phasing out of separate development and power sharing with blacks, would have served only
to legitimise the South African apartheid regime.
Finally, dialogue would have also undermined UN efforts
in trying to bring about meaningful change in South Africa.

By contrast to the uncertainties and contradictions
of dialogue, armed struggle as an appropriate method to
effect political change in Southern Africa had been given
credence through the coup d'etat that toppled the
colonialist government of Portugal. The coup in Portugal,
partly inspired by the effects of the colonial wars, led
directly to the independence of the Lusophone states:
Angola, Guinea-Bissau, and Mozambique. In addition, the
coup weakened the capacity of the white minority regime
in Rhodesia to resist guerrilla and diplomatic pressures
wielded by nationalist forces in the territory. This
unfavourable strategic situation from the mid-1970s onwards prompted South Africa's frantic efforts to blunt
the possible adverse effects of an independent, militant
Zimbabwe under indigenous and genuine majority rule.
South Africa knew that a black-ruled Zimbabwe would
(1) shrink the South African defence perimeter; (2) give
moral, diplomatic and, in the long run, logistical support
to the opponents of the regime; and (3) be a curtain-
raiser to the fate of the apartheid regime itself.

To obviate these eventualities, John Vorster, then
Prime Minister of South Africa, advocated detente with
neighbouring states. His major motives were (1) to contain radical, seemingly "Marxist" regimes in Angola and
Mozambique; (2) to placate other Front Line States such
as Botswana, Tanzania, and Zambia; and (3) to slow down
or forestall nationalist attacks on South Africa. In
this way, Vorster hoped to buy time for apartheid in
the Republic of South Africa as well as a secure place
for the white minority in Rhodesia.

African reactions to the detente initiative, as to
the dialogue attempts, were largely negative. Whilst
the Front Line States, especially Zambia, were initially
in favour of detente, the OAU Council of Ministers roundly
and firmly condemned it through the Dar-es-Salaam
Declaration. This Declaration was fully endorsed by
the 1975 OAU Kampala summit, which described detente
as an attempt to "legitimise the oppression and exploitation of the South African people,"[31] although it left
open the door for the negotiation of the independence of
Rhodesia and Namibia.

In any case, detente with South Africa could not have succeeded for the following reasons. First, the South African government has not relented in its suppression of blacks as exemplified by its massive repression and murder of protesting African students in Soweto in 1976. Second, its proclamation of "good neighbourliness" notwithstanding, the South African government had demonstrated its aggressive policy towards its neighbours by the invasion of Angola in 1975 and subsequent raids into and, eventual occupation of the southern portion of, that country in pursuit of SWAPO guerrillas. And, third, the possession of nuclear capability by South Africa with the assistance of Western states had given the lie to Pretoria's proclaimed peaceful intentions in the Southern African sub-region in particular, and the entire continent in general.

South Africa's frenzied attempts to develop a nuclear bomb are meant (1) to preempt direct military confrontation by African states who in any case had no such capability and (2) to blackmail the OAU states to acquiesce in apartheid. Any South African nuclear capacity is, however, largely irrelevant given the fact that there is no way in which it could be used without affecting the white population. Related to this is the fact that a nuclear bomb is not a suitable weapon in a guerrilla war in which the ANC, with its strategic attacks against key South African government installations, has been waging. The OAU endorses these attacks against South African targets and has provided the necessary diplomatic support, which has facilitated the training of freedom fighters.

What are the immediate prospects for the guerrilla strategy in South Africa? First, the guerrilla attacks--intermittent and at times uncoordinated--will not bring about black majority rule overnight. The South African government is well equipped to fight a counter-insurgency war and has a wealth of experience of "counter-insurgency" based on its assistance to the former Rhodesian government of Ian Smith. However, guerrilla warfare is only one phase in the long struggle. With the increased politicisation and mobilisation of Africans--through school boycotts, strikes, and industrial sabotage--the apartheid administration may yet be worn down so much as to lead to its internal collapse. The OAU's role in these events, designed to make apartheid crumble through its own contradictions and vulnerabilities would not be so different from what it has been in the past. However, once Namibia is free, the Organisation will be able to concentrate exclusively on the white redoubt: South Africa itself.

NOTES

1.   CIAS/Plen.2/Rev. 2.

2. James Robertson, *Transition in Africa* (London: Hurst, 1974), p. 250.

3. *West Africa*: 3 June 1958.

4. OAU Co-ordinating Committee for the Liberation of Africa, Minutes of the Second Session (New York, October 1963), p. 6.

5. Colin Legum, *Pan-Africanism* (London: Pall Mall, 1962), p. 254.

6. See Chapter 1 for more details.

7. See News Letter by Movement for Justice in Africa (MOJA) (Monrovia, September 1973).

8. Figures compiled from the following sources: International Monetary Fund, *Direction of Trade: A Supplement to International Financial Statistics Annual 1961-65* (Washington, D.C.: Statistics Bureau of the IMF, nd.), p. 214; also the issue 1966-70, p. 314.

9. OAU, Council of Ministers' Resolution, CM/Res. (4)1.

10. LC 21/DOC. 5, p. 213.

11. *Africa Research Bulletin* (Political Series), May 1975, p. 3618.

12. LC 36/DOC. 5, Annex VIII.

13. LC 21/DOC. 5, p. 214.

14. Although we are not unaware of the OAU's role in Mozambique, we do believe, however, that it was much more active in Guinea-Bissau and Angola. We have accordingly given priority to the two territories in this chapter.

15. *Africa Contemporary Record*, Vol. 8, 1975-76, p. A7.

16. *New York Times*, 13 January 1976.

17. CIAS/Plen.2/Rev. 2.

18. ECM/Res.14(11) 1964.

19. ECM/Res.13(vi). The states that severed relations were Algeria, Congo-Brazzaville, Ghana, Guinea, Mali, Mauritania, Sudan, Tanzania, and United Arab Republic.

20. See Berhanykun Andemicael, *The OAU and the UN* (New York: Africana, 1976), p. 115.

21. Zdenek Cervenka, *The Unfinished Quest for Unity* (New York: Africana, 1977), p. 125.

22. See Amadu Sesay, "The Roles of the Front Line States in Southern Africa" in Olajide Aluko and Tim Shaw (eds.), *Southern Africa in the 1980s* (London: Allen and Unwin, 1983).

23. *Africa Research Bulletin* (Political Series) September 1976, p. 4168; and *West Africa*, 11 October 1976, p. 1507, and 1 November 1976, p. 1611.

24. See Richard Bissell, *Apartheid and International Organizations* (Boulder: Westview, 1977).

25. OAU, CM/Res. 87 (vii).

26. For details, see Thomas Gallaghy, "Zaire and Southern Africa," in Aluko and Shaw (eds.), *Southern Africa in the 1980s*.

34

27. Africa Research Bulletin (Political Series)
February 1977, p. 4311.
28. The Military Balance 1971-72 (London: IISS,
1972), p. 39.
29. For details of strategy, see Lusaka Manifesto
(Lusaka: Government Printer, 1968).
30. OAU: The Principles of the OAU Charter, the
Lusaka Manifesto, Dialogue and Future Strategy (Addis
Ababa: OAU Secretariat, June 1971).
31. Colin Legum (ed.) Africa Contemporary Record
(London: Rex Collings, 1976), Vol. 8, p. A70. See also
Africa Research Bulletin, April and August 1975, pp. 3583-
86 and pp. 3719-22, respectively.

3
The OAU and African Conflicts

Both international and regional organisations state
as one of their objectives not only the prevention of
conflicts but also the pacific settlement of any disputes
among member states. The proclamation of peace as the
guiding hand in the relations among states is necessi-
tated by their desire to curb the use of force in
settling disputes. In addition, it is meant to restrain
powerful states from preying on weaker members as African
countries become more unequal, the propensity for such
intra-continental intervention will grow.
The covenants of global and regional organisations
are a reflection of the circumstances prevailing before
and during their formation. In the case of the OAU, the
African environment was characterised by various types of
conflicts. These conflicts fall into three broad categories: (1) challenges to the integrity of the state,
i.e., secession, territorial and boundary claims; (2)
challenges to the integrity of the regime, namely,
assassination plots, subversion both internal as well as
external; and finally (3) ideological and personality
disputes. Concern over this range of conflicts is
reflected in several articles of OAU Charter. The core
Charter provision on conflict is Article III which calls
for, among other things,

> non-interference in the internal affairs of
> States; respect for the sovereignty and terri-
> torial integrity of each State and for its
> inalienable right to independent existence;
> peaceful settlement of disputes by negotiation,
> mediation, conciliation or arbitration, un-
> reserved condemnation, in all its forms of
> political assassination as well as of sub-
> versive activities on the part of neighbouring
> States or any other State.

A major objective of the founding fathers in
Article III is the maintenance of the political and

territorial *status quo* inherited by the new ruling elites
of Africa. These elites were cognisant of the fact that
revision of the European-imposed boundaries and challenges
to the legitimacy of their regimes could open up a
pandora's box. In short, in the interest of political
stability, peace, and social and economic development in
Africa, the new states must endeavour to settle their
intra-continental disputes peacefully.

However, the OAU did not formally institutionalise
its conflict control machinery until 1964 at the Cairo
summit. At that meeting, the Organisation approved a
special protocol setting up a Commission of Mediation,
Conciliation, and Arbitration. The Commission was to have
twenty-one members to be elected by the Heads of State for
a period of five years. In spite of elaborate techniques
for conflict control, the Commission was a dismal failure.[1] Indeed, before it was formally inaugurated in 1965,
the OAU had already evolved other conflict control methods.

METHODS OF CONFLICT CONTROL BY THE OAU

The OAU has used a variety of techniques to control
conflict among its members. For analytical convenience,
these techniques can be grouped into two categories:
formal and informal. Both categories are political rather
than legal, economic, or coercive. Under the formal
techniques of conflict control are the following sub-
categories: (a) conference or summit diplomacy; (b) ad
hoc committees; and finally, (c) good offices committees.
However, the only informal method of conflict control is
"presidential mediation." This is a unilateral initiative
of particular African presidents which might later receive OAU approval; in some cases it might even be
incorporated into formal institutional channels.

Conference Diplomacy

The summits of the OAU bring together authoritative
decision-makers of the Organisation. As such, the summits
offer the most competent and highest political forum to
discuss, treat and sometimes resolve outstanding political
issues. Like the annual meetings of the Heads of State,
the Council of Ministers meetings are also used in part
to seek peaceful solutions to disputes. States which are
a party to conflicts or their "allies" have from time to
time initiated moves either at the regular or extraordinary meetings of both bodies with the view to resolving such conflicts within an African context.

Conference diplomacy is not an entirely imaginative
or effective method for settling conflicts, although it
is routinely used by regional as well as global organisations. It has several weaknesses. One such weakness is

its openness which allows conflicting states to use the occasion for propaganda and pressure. The debates in such a forum may be used to embarrass the other party rather than to seek an amicable solution to the dispute. This argument notwithstanding, there are still some advantages to conference diplomacy. First, it sets the context for pacific settlement where the parties are pre-disposed to a solution. Second, summit diplomacy--because of the elaborate arrangements surrounding it--sometimes causes delay and so allows tempers to cool down. This is an indirect way of putting the issue on ice, although such "icing" is not a substitute for a permanent solution. In such circumstances, change in leadership or capability in any of the conflicting states may lead to the reactivation of the dispute. This was the case with the Ethiopia-Somalia conflict over the Ogaden, which had remained dormant roughly between 1969 and 1977 when it erupted into a full scale war between the two countries.

Ad Hoc Committees

The establishment of ad hoc committees comprising selected OAU Heads of State or their representatives is another distinctively African technique of conflict control. This method is sometimes used concurrently with conference diplomacy. There are at least three criteria for selection onto an ad hoc committee. These are: (1) perceived neutrality of a state; (2) chairmanship of the OAU; and (3) nearness of a state to the area of conflict.

A variant of the ad hoc committee system is the Consultative Mission. However, this has been used only once, during the Nigerian civil war. The objective then was to reconcile the Federal Nigerian Government and the secessionists. Although the six-member panel--Cameroon, Ethiopia, Ghana, Liberia, Niger, and Zaire--succeeded in bringing the combatants to the conference table, it was not able to get them to accept cessation of hostilities. The secessionists were eventually overwhelmed by the stronger military force of the Federal Government in January 1970.

Good Offices Committees

The final category of conflict resolution is the use of the "Good Offices" of African Chiefs of States. This is an indirect way whereby the OAU brings its influence and/or pressure to bear on conflicting parties in the hope of containing or resolving a dispute. The "Good Offices" technique seeks to bring together interested parties to a local conflict. There are two broad

criteria for selection into a "Good Offices" committee,
namely the perceived influence of a state and the standing of an African leader among his colleagues. Nigeria
has served on many such "Good Offices" committees on the
basis of the first criterion while Emperor Haile Selassie
of Ethiopia and President William Tubman of Liberia were
often chosen as members of "Good Offices" committees on
the basis of their personal prestige. Nigeria was, for
example, the Chairman of the "Good Offices" committee on
the Ethiopia-Somalia dispute.

Presidential Mediation

     Presidential mediation is closely related to the
"Good Offices" technique. There are, however, important
differences between the two. Whilst "Presidential Mediation" is an essentially unilateral initiative of an
African Head of State acting as a disinterested mediator,
a "Good Offices" procedure, on the other hand, is
exclusively initiated by the OAU. As noted earlier, however, "Presidential Mediation" can be and often is
legitimised later by the Organisation. The Algeria-
Morocco dispute of 1964 can be considered as the first
and classical case of "Presidential Mediation." During
the dispute, the late President Modibo Keita of Mali and
Emperor Haile Selassie intervened and succeeded in
securing a ceasefire in the border war between the two
countries. It was only after the ceasefire had been
arranged that the OAU gave formal support to their peace
initiative.
     "Presidential Mediation" has been widely used to
contain African disputes. This is explicable for a couple
of reasons. First, the longevity of its existence; it
predated the OAU. And second, most African conflicts are
of a political nature; consequently disputants have come
to rely on African leaders to find political solutions
to such disputes.

RESOURCES OF THE OAU FOR CONFLICT CONTROL

     Oran Young has identified certain attributes which
regional and international mediators must possess if they
are to control effectively disputes among an institution's
members.[2] These attributes can be classified into two
broad categories: material and ideal resources. Material
resources include such indices as the quality of the
personnel in the organisation. By this is meant their
level of training, skill and knowledge of politico-
military affairs. The mediating organisation must be
able to finance and equip its personnel to enable them to
perform an interventionary role in a conflict situation.

On the other hand, ideal resources are in the main concerned with the impartiality of an organisation in a
conflict situation; that is, the mediator must be seen by
the disputants to be neutral. As well, the institutions
should be considered important in terms of being the
appropriate forum for a discussion of issues in dispute.

Using the above criteria one can conclude that the
OAU is a low resource organisation. For example, the
Organisation's budget for 1980 was a mere $17.6 million,
hardly enough to run the Secretariat very effectively.
One striking feature about the OAU budget is the high rate
of default which Secretary Generals have annually
complained about. Edem Kodjo lamented in 1980 that over
$9 million was outstanding as membership contributions to
the Organisation; he was constrained to call "for the
application of the measures (which he did not elaborate)
specified by the new financial text."[3]

Perhaps the most dramatic illustration of the OAU's
low-resource capability is its inability to establish an
African High Command despite broad agreement on the need
for such a force since 1964. Such a High Command, either
at the regional or sub-regional levels, could have helped
to contain disputes. However, finance is not the only
stumbling block against the formation of such a force,
that could also be used to deter as well as contain
aggression. There are many other reasons for the failure
of the OAU to establish the High Command. These are the
lack of political will on the part of many African states,
differing ideological and socio-political systems, and the
obstinate attachment of African leaders to the idea of
state sovereignty.[4]

Apart from the absence of a High Command, the OAU
also lacks diplomatic and technical skills that could be
useful in peacekeeping activities. Although African
commanders have in the past headed United Nations Forces
in the Congo and Lebanon, it is doubtful whether this
experience could be of much use to the OAU. It should be
pointed out that the logistical support enjoyed by these
commanders in those circumstances has been provided by the
developed countries. Besides, the United Nations is much
better endowed with financial and other resources than
the OAU. The OAU's logistical handicap has been vividly
illustrated by the Chad fiasco examined in detail below.

Unlike the UN, where Secretary-Generals are world-
renowned diplomats, OAU Secretaries have generally been
inexperienced without independent reputation except that
attaching to their governments. Besides this major
disadvantage, there is a high rate of turnover in the
office of the Secretary-General. Except for the first
Secretary--Diallo Telli of Guinea--who occupied the post
from 1964 to 1972, all other Secretaries have had a short
stint at the Secretariat. The rapid turnover has made it
difficult for incumbents to acquire the expertise and
prestige that should go with the office. Long periods in

office, with the development of accompanying diplomatic
skills might have compensated for the absence in the OAU
Charter of specific responsibility for conflict control by
the Secretary-General. In the circumstances, the OAU
chief executive has been denied the power of effective
independent initiative in conflict control; hence the
frequency of presidential involvement.

The OAU appears to be well endowed with the ideal
resources of salience, prestige, membership expectation,
and the desire that it should continue to exist. It is
the only regional Organisation which brings together the
different ideological, cultural and racial groups in the
continent. As such, it is the Continental "melting pot."
This notwithstanding, the OAU has on several occasions
disappointed even its most ardent supporters by its in-
decision on crucial issues.

We shall in the rest of the chapter examine the OAU's
role in four disputes. These are the Chad civil war,
the Ethiopia-Somali conflict, the Tanzania-Uganda
imbroglio and the Western Sahara impasse. The cases are
chosen for a variety of reasons. First of all, all four
disputes fall within the broad conflict types categorised
above. Second, some of the issues span the lifetime of
the Organisation. And third, some other issues impinge
on certain principles of the OAU such as the idea of non-
intervention and self-determination.

THE OAU AND AFRICAN DISPUTES

The Chad Civil War

The recent Chad conflict was in many ways like the
Congo (now Zaire) conflict of the early 1960s. It was
initially a challenge to the integrity of a governing
regime which subsequently became a civil war between
various factions. The OAU as a rule, does not interfere
in the internal affairs of its members, although one must
hasten to add that the two significant exceptions were the
Nigerian and Congolese civil wars. The Organisation did
not, for example, recognise the civil wars in Sudan and
Burundi. Equally, it maintained an embarrassing silence
on the Eritrean secessionist war in Ethiopia.

The conflict in ethnically and racially variegated
Chad had two inter-related dimensions. In the first
place, it was a case of internal strife between ethnic
and political factions. And second, compounding the
strife was the intervention by many other state actors
who maintained a clientele relationship with the various
armed groups in the country. Foremost among the
intervenors in the civil war was France, the country's
erstwhile colonial master. Libya also intervened to
bolster whatever armed fraction would condone, if not
approve, its annexation of the mineral rich Aozou strip.

The Chad conflict brought into bold and clear relief the economic, religious and ethnic differences in the country; it pits Southern Chad, predominantly Christian and animist, against an Arab and Islamic North. The predominance, and control of the government as well as the administrative machinery of Chad by Southerners, mainly Sara, was resented by the northerners who were neither westernised nor literate in terms of western script. A northern guerrilla organisation--the Front for the Liberation of Chad (FROLINAT)--which later broke into many factions, took up arms against the central government of President Tombalbaye in 1967 and intensified the war against his military successor, General Malloum.

The Chad problem was first tabled before the OAU in 1977 when General Malloum complained to the Assembly about Libyan intervention in the country through Gadaffy's support for various FROLINAT forces fighting his government. The OAU first employed its more popular technique of mediation; it established an Ad Hoc Committee of six states--Algeria, Cameroon, Gabon, Mozambique, Nigeria, and Senegal--to investigate the allegations of Malloum. But the Committee was ineffective in that it could convince neither the combatants to lay down their arms nor the Libyans to withdraw. Similarly, it failed to convince the guerrillas to come to the peace table with the government of Malloum.

It was probably the inactivity of the OAU Ad Hoc Committee that prompted individual African states, mainly the neighbours of Chad, to initiate efforts in the form of "Presidential Mediation," which was begun by Colonel Gaddafy of Libya. This successor effort to the earlier ad hoc attempt did not produce any significant result until August 1978 when, as a result of the Khartoum agreement, Hissene Habre, leader of the break-away faction of FROLINAT--Armed Forces of the North (FAN)--joined the government of Malloum as Prime Minister. However, the reconciliation between Malloum and Habre was short-lived, as fighting broke out again in February 1979.

Nigeria, Chad's most powerful and influential neighbour, was approached in 1979 separately by France and Sudan as well as leaders of the strife torn country, to convene a conference of the various factions in Chad. Over a period of six months between March and August 1979, Nigeria with other neighbouring states hosted four "peace conferences" on Chad.[5] Although it succeeded initially in producing an accord between the various fractions of FROLINAT and the central government, because of its own limited material resources, Nigeria could not sustain or effectively coerce the combatants to honour the agreements reached in Kano and Lagos. A Nigerian peace force sent to Chad to maintain order was not able to separate the myriad factions in the country. Indeed, it was withdrawn from Chad in June 1979, when disagreement arose between Lagos and the Mohamed Shawa regime in Ndjamena, which had been formed in violation of the Kano Accords. Nigeria subsequently imposed economic sanctions on Chad by withholding oil exports to that country. In addition, the Nigerian government blocked the Shawa regime from being seated at the 1979 OAU summit in Monrovia.

The role of the Organisation during the Nigerian mediation exercise was that of legitimator of the on-going peace talks. At the 1979 summit, the OAU commended Nigeria and urged its leaders to continue their search for peace in Chad. The OAU similarly legitimised the mediation efforts of Togo which has even less material and ideal resources than Nigeria. Togo's mediation followed the outbreak of war between the forces of President Goukhouni Weddeye and those of his Defence Minister, Hissene Habre, in March 1980. Accompanied by the Organisation's Secretary General, Edem Kodjo, Togolese President Eyadema was able to secure an agreement for a truce between the warring factions. However, this truce lasted only two days. And all efforts by the OAU to stop the fighting failed. In the event Weddeye turned to Gaddafy for military and political support while Habre secured assistance from Egypt and Sudan. The military intervention of Libya in December 1980 led to the defeat of Habre's Egyptian- and Sudanese-backed forces.

The opposition to Gaddafy became more intense in January 1981 following the announcement of a (misunder-stood) Chad-Libya agreement, which many states in the region as well as France and the United States of America had interpreted as a proposed merger between the two states. The misperceived view of "Libyan imperialism" in the West African Sahel region created an alliance of conservative/moderate African states, particularly Senegal, Sudan, and Nigeria, with strong French backing, to demand the withdrawal of Libya from Chad and its replacement by multilateral OAU Peacekeeping Force.

This rather anti-Libyan position was endorsed by the OAU summit of June 1981 in Nairobi, Kenya. Although the Organisation did not explicitly call on Libya to with-draw, it nevertheless implied this when it revived the idea of a peacekeeping force to be sent to Chad (such a force would, of course, displace the Libyans). In addition, the OAU weakly called on Chad's neighbours not to engage in "acts of destabilisation or sabotage against Chad."[6] The call was hardly likely to be heeded given the conflicting and multiple indigenous interests present in Chad, which openly sought the assistance of external patrons to further their particularist interests in the country.

The OAU Peace Keeping Force, which in the case of Chad had been mooted since 1979, did not take off until December 1981 and then in circumstances which compromised the autonomy of the Organisation and exposed the dependence syndrome of African states. A myriad of intra- and extra-continental interests was responsible for the creation of the OAU Peace Force.

The initiative to translate OAU wishes about the Force to the actual establishment of that Force was made by the newly elected social democratic president of France, Francois Mitterrand. President Mitterrand's suggestions had the diplomatic support of the United States government which for its own reasons wanted Libya out of Chad. Nigeria, which was probably consulted by France before Mitterrand's suggestion for the creation of the Force was made public, said it would contribute troops if the OAU requested it to do so.[7]

The OAU had requested six states representing radical as well as conservative/moderate opinion on the continent to contribute to the proposed Peace-Keeping Force for Chad. There were seemingly radical Benin and Guinea and the conservative/moderate states of Nigeria, Senegal, Togo, and Zaire. However, only three conservative states actually supplied troops--3,000 of them, about half of what the OAU had originally requested. Moreover, the three states which supplied troops for the Peace-Keeping Force relied on Western states for logistical as well as financial support for their troops in Chad. Such reliance by Senegal and Zaire for logistical and financial assistance from the U.S. and Great Britain seriously compromised not only the autonomy of the OAU operations in Chad but also tainted the image of the Organisation as a tool of Western capitalist states.[8]

The formation of the OAU Peace-Keeping Force under Nigerian command and the deployment of the force to Chad to keep the warring factions apart might, at first sight, appear to be a significant milestone for the Organisation in its attempts to control conflicts among and within its members. However, as previously noted, the origin and composition as well as the financing of the force left the question of its neutrality much in doubt. This was further compounded by the ambiguity of the mandate given to it.

In line with the general mandate of UN Peace-Keeping Forces, the OAU construed the role of the force in Chad as that of a neutral arbiter between the loosely grouped forces of GUNT and the advancing forces of Habre's *Armees du Nord*. As OAU Secretary-General Kodjo was quoted as saying, the force is "to maintain peace and safeguard Chad's security without fighting against other factions."[9] This had been the opinion of the OAU representative in Chad who in November said that the Organisation did not intend to repeat the United Nations experience in the Congo where the world body collaborated with the government of that country to defeat a rebellion.[10]

OAU neutrality, which was more apparent than real-- the Organisation had given tacit support to Weddeye and his GUNT by accepting him into its meetings--was not what President Weddeye wanted. He wanted the OAU to be an extension of his broad, tenuous and at times antagonistic coalition government which would fill the role as well as the vacuum created by the departure of Libyan forces. The reluctance of the OAU forces to stop the advance of Habre, thereby assisting Weddeye, led the latter to threaten that he would call on his friends for military assistance.

The disputed mandate of the OAU force created a political problem for that force and the GUNT Government in Ndjamena. A more daunting problem though, in the form of finance, confronted the peacekeepers. In February 1982 the OAU Committee on Chad (Central African Republic, Guinea, Kenya, Nigeria, and Zaire) met in Nairobi to discuss the maintenance of the financial problem of the Force and set up an unprecedented political schedule for the settlement of the civil war. By establishing such a time-table, which, *inter alia*, called for a ceasefire, negotiations, and the drafting of a constitution to be followed by elections in the territory, the Organisation had brushed aside its sacrosanct principle of non-interference in the internal affairs of a member state.

Subsequent OAU resolutions on the Chad civil war represented a significant shift from the Organisation's previous position of endorsing the Government of Weddeye to putting Weddeye and Habre on the same footing. This was not lost on the Chadian leader. A relevant question, then, is why did the OAU shift its position? There are two possible reasons for this, and one of them is related to the Organisation's poor material resource base. The first reason is the Organisation's failure to raise the staggering $163 million which Secretary General Kodjo said would be needed for the upkeep of the peacekeeping force. And second, the military defeats of the GUNT forces of Weddeye convinced the OAU that Habre is a viable force that should be reckoned with in any settlement of the Chad crisis.

That a political solution could be found to two decades of Chadian civil strife within four months of OAU deadline for the withdrawal of its forces seemed improbable. The various factions in Chad and their divisive client patrons outside seemed bent on continuing the war until they obtained a solution satisfactory to themselves. Consequently, 15 March 1982, which was to have been the day negotiations were to begin after a 1 March ceasefire--which was never respected by the warring factions--was just another day of fighting in Chad. The failure of the OAU proposed ceasefire to take effect laid to rest its whole political agenda for Chad. On 7 June 1982 Habre's forces captured Ndjamena and the OAU peace-keeping force, which was scheduled to withdraw at the end of that month, began pulling out three months before the expiration of its mandate.

The victory of Habre in a complex situation like the one in Chad, as would be expected, has not led to a lasting solution to the conflict. Soon after his flight from Ndjamena, Weddeye reorganised his forces and launched a counter-attack. By July 1983 his forces effectively controlled the northern third of the country. Worse still is the unprecedented internationalisation of the conflict, which saw massive French and American intervention on the side of Habre.

At this juncture, an evaluation of the performance of the OAU Force in Chad is important for the light it may shed on the future of such forces raised by the Organisation. The overwhelming consensus in Africa is that the OAU peacekeeping force failed to keep the peace in Chad. There are numerous reasons why this is so. The first was the absence of logistics for the troops. As noted earlier, the states which contributed troops for the operation in Chad had to rely on France, the United States, and Great Britain for logistical support. Closely related to this point is the second one, that is, the grave financial problem which beset the OAU Force. Again, the money and upkeep of the Senegalese and Zairean forces was provided by France and the United States. As the Commander of the Forces opined, "The operation is that of the OAU only in name. The countries that supplied the troops were financially responsible for them," and he concluded that the OAU was incapable of sponsoring a force "for the maintenance of peace in Chad."[11] Indeed, the OAU has so far made no financial reimbursements to the states that contributed to the peace force. The combination of the first and second reasons led to the third; the late arrival of the troops in Chad. The Force arrived in Ndjamena in the middle of December, four weeks after the Libyan withdrawal. The gap between the arrival of the troops and the departure of the Libyan forces was exploited by Hissene Habre who launched new offensives against GUNT. And even after the OAU Force had arrived in Chad, Habre's triumphant march into Ndjamena continued since the Force could not engage his troops in combat. It therefore watched helplessly while Habre advanced steadily towards the Chadian capital. This led to his victorious entry on 7 June and the flight of Weddeye.

A fourth reason was the controversy over the mandate of the Force. As we have noted earlier, the OAU wanted its force to play the role of UN forces, to act as a buffer between the warring factions. But it was ill equipped even for this limited role. On the other hand, Weddeye wanted the Force to help his GUNT to flush out Habre and his rival forces from the country. There was never to be an agreement over the role which the Force was to play. Besides, there was a lot of interference with the command structure of the Force from national governments.[12] This situation, coupled with the lack of equipment and food shortages, led to the demoralisation of the troops.

The fifth reason for the failure of the Peace-Keeping Force concerns the lack of support from Chad's neighbours as well as the failure to enforce a ceasefire in the territory. Thus, while the Force was in Chad some OAU members, particularly Sudan, gave active support and encouragement to the forces of Habre. This support enabled him to secure a platform from which to launch his successful counter-offensive against Weddeye's GUNT. And a final reason is the "conspiracy" theory. According to this theory, the OAU actively colluded with Western imperialist powers to drive out Weddeye and his ostensibly pro-Libyan government from Ndjamena. In his place, the Organisation and the West encouraged the enthronement of Habre who had all along been receiving generous material, financial, political, and indeed, moral support from the West through their proxies--the conservative Arab states, particularly Egypt, Saudi Arabia, and Sudan-- which are patently anti-Libya. This collusion is evident even in the very composition of the OAU Force. As noted earlier, the troops came from three conservative and pro-Western regimes: Nigeria, Senegal, and Zaire. Besides, the Senegalese and Zairean troops were, as already stated, financed by the U.S. and France, both leading western countries.

Lessons for the future can clearly be drawn from the above set of reasons. First, the OAU would, for the foreseeable future, never be able to launch a successful peacekeeping force unless it is able to finance it as well as provide it with the necessary logistical support locally. Second, the Organisation should secure a cease-fire before it embarks on peacekeeping activities in any troubled spot on the continent. This would prevent the Chad experience where Habre's forces were able to exploit the fact that there was no ceasefire to consolidate their own position. Third, the OAU should get a clear mandate for its force before it is sent out into the field. This would avoid either lingering doubts or high expectations on the part of the authorities in the state concerned, about whether the force would or would not fight on their side. And finally, the Organisation must ensure in future that member states do not aid dissident forces while peacekeeping activities are going on. If the above suggestions are heeded, then, the OAU should yet be able to launch a successful peacekeeping activity on the continent. This, of course, also assumes that the Organisation would have enough funds to maintain the force. Such funds should be generated within the continent as that is the only way that could give the OAU the necessary autonomy and authority both of which it lacks at the moment.

The Ethiopia-Somalia Dispute

The Ethiopia-Somalia dispute involves a threat to the integrity of a state: Ethiopia, a member of the OAU. Somalia has been contesting a part of Ethiopian territory known as the Ogaden, which is inhabited by people of Somali stock. However, although the principle of self-determination is central to the Charter of the OAU, it only sanctions self-determination for peoples in non-self-governing territories or those which have been dominated by white supremacist regimes. Accordingly, self-determination for ethnic or national groups within black-ruled states is excluded. Thus, when the Somali delegation to the inaugural conference of May 1963 called for "self-determination of the inhabitants of the Somali areas adjacent to the Somali Republic," the resolution was summarily dismissed by both the Foreign Ministers and the Heads of State. In trying to find a peaceful solution to the dispute, however, the OAU has employed two modes of conflict resolution: Conference Diplomacy/Summitry and Good Offices Committees.

In 1964, when Somalia and Ethiopia were in open conflict, the issue was discussed by an Emergency Council of Ministers (ECM) held in Dar-es-Salaam from 12 to 15 February. Although the ECM was convened to discuss mutinies in East Africa,[13] the grave situation created both for the OAU as an Organisation and for Africa as a whole meant that the Conference could not skip over the question of the Ogaden. This is because, as we have already emphasized, the heart of the conflict was a serious challenge to one of the cardinal principles of the OAU as well as to one of the rules of intra-African relations. A successful challenge to the territorial integrity of Ethiopia could have meant the demise of the new Organisation because the upholding of this principle was perhaps the major incentive for most African states to retain their membership. The Council of Ministers passed a resolution calling for an immediate cessation of provocation and insults by both sides, and articulating the guidelines that the dispute should be settled "in the spirit of paragraph 4 of Article III of the Charter." The Somalis found this unacceptable because in fact the resolution legitimised the *status quo ante*--that is, that Ethiopia would remain sovereign in the Ogaden.

At the Ordinary Meeting of the Council of Ministers in Lagos which came on the heels of the Emergency meeting,[14] a more anti-Somali resolution was passed, this one mentioning Article III, paragraph 3, and thus reaffirming the OAU's commitment to respect and defend Ethiopia's territorial integrity. Hostilities ensued immediately. President Aboud of Sudan was authorised by the OAU to mediate and in March 1964 the Foreign Ministers agreed on a ceasefire, a withdrawal of military

forces from either side of the border, and the resumption
of negotiations in pursuance of the resolution adopted in
Lagos. The Somalis were forced to accept this position
because they were no match for Haile Selassie's well-
equipped army, and Somalia could get no diplomatic or
military support anywhere in the world for its war aims.
But the Somalis still would not give up their hope of re-
gaining the Ogaden from Ethiopia. After the OAU un-
equivocally rejected their claims in its Declaration on
Boundaries at the 1964 summit in Cairo, the Somali
National Assembly unanimously passed a resolution declar-
ing that the Cairo OAU summit Resolution "should not bind
the Somali Government."

Between 1965 and 1967, the Ethopia-Somalia border
remained relatively peaceful. In 1967 Mohammed Cigaal
became Prime Minister of Somalia. He pursued a policy of
peaceful co-existence and detente with his two neighbours
--Kenya and Ethiopia--on whom his country had irredentia
claims. The result of this policy was that Mohammed
halted the supply lines to Somali guerrillas operating in
Ethiopia. That same year, the Somali leader and Emperor
Haile Selassie of Ethiopia agreed to a joint military
commission which was to monitor their common border.
Somalia also agreed to stop hostile propaganda against
Ethiopia.

But the period of detente lasted only two years. In
1969, the government of Cigaal was toppled in a *coup
d'etat* led by Mohammed Siad Barre. Although Siad Barre
wanted to reunify the Somali lands, the first few years
of military administration were spent on the building
of socialism in Somalia. Accordingly, the new military
authorities counselled patience and peace in regard to
the dispute with Ethiopia. At the same time, however,
Barre was building up a military capability with the help
of the Soviet Union which was to be used eventually in
Somalia's bid to recover the Ogaden.

In 1973, in line with his commitment to reunify the
Somali lands, President Siad Barre and Foreign Minister
Omar Arteh undertook a diplomatic offensive. In the May
1973 OAU Council of Ministers meeting, Somalia tried to
place "its territorial dispute with Ethiopia" on the
agenda. Ethiopia objected as the Somali position, it
claimed, was "contrary to the OAU Charter." There was a
stalemate, as the Somalis had the support of the North
African Arab states while Ethiopia had the support of all
sub-Saharan Africa. To avoid a bitter confrontation, the
Nigerian Foreign Minister, Okoi Arikpo, engineered a five-
man commission, which he would chair, to be composed of
the foreign ministers of Algeria, Ivory Coast, Liberia,
and Tanzania. But when the Commission was preparing to
put the dispute before the Heads of State summit, Somali
Foreign Minister Arteh, reading the mood of the delegates,
declared that his country did not plan to bring up the
dispute again before the Heads of State Assembly. In

fact, President Barre, who claimed an imminent Ethiopian attack on Somalia, initially refused to attend the summit. But Chairman Gowon of Nigeria sent him a special invitation and during the summit an eight-man Good Offices Committee was established.

The Committee empowered Presidents Numeiry of Sudan and Gowon of Nigeria to explore the possibility of resolving the conflict. Gowon's plan, modelled on the Algeria-Morocco border pact, called for the joint development of the Ogaden in the context of Ethiopian sovereignty. However ingenious this proposal was, it did not satisfy the Somalis. In 1974, when Siad Barre was Chairman of the OAU Heads of State meeting, he tried to convince his colleagues of the Somali cause. Selassie, whose position in Addis Ababa was becoming tenuous, at first remained in Ethiopia. Although Gowon was able to persuade Selassie to attend, he could report no progress on his peace proposal. He said that only the two countries could find a satisfactory solution to the conflict.

In September 1974 Selassie was overthrown in a military coup. In the aftermath of the revolution, there were a lot of separatist agitations against the Amharic dominated Ethiopian State. However, Ethiopia's border with Somalia remained quiet. There are a number of reasons for this. First, as Chairman of the OAU, the Somali leader did not want to be seen openly to be flouting one of the cornerstones of the Organisation by engaging Ethiopia in an irredentist war. And second, the Soviet Union, upon which the Somalis depended for military hardware, was opposed to the irredentist claims and pressured Siad Barre against military confrontation with Ethiopia.

However, by 1977 regular Somali troops in the guise of Ogaden guerrillas marched into the Ogaden. Early substantial victories by Somali forces in August forced Ethiopia to seek external aid from the OAU. In response, President Bongo of Gabon, then Chairman of the Organisation, attempted to resuscitate the Good Offices Committee to consider the "Somali invasion" which Ethiopia requested. This was, however, impossible because as noted earlier, the Somali victories in the field led them to snub any offer of conciliation by the OAU. The fighting therefore continued.

In 1978 an OAU peace plan, which among other things provided for a six-mile demilitarised zone along the border and an end to Ethiopian air attacks on the Somalis, was categorically rejected by the Ethiopians. In rejecting these peace proposals, the Ethiopians argued that they did not take "account of the guerrilla war in the Ogaden waged by Somali regular troops dressed as nomads." In fact, whatever the reasons for rejecting the peace efforts of the OAU, it was clear that the tide of the war was turning slowly but surely, in favour of the Ethiopians who were now receiving help from the Soviet

Union. Thus they had no incentive to negotiate from a
position of strength. At that stage the OAU could not
play any effective mediatory role in the conflict. The
war therefore continued unabated until 1979, when the
Ethiopians, using superior firepower, decisively defeated
the Somalis. The decisive victory of Ethiopia forced the
Somalis to sue for peace. Thereafter the border again
remained relatively quiet. However, at the time of
writing, there have been reports of renewed war between
Ethiopia and Somalia over the Ogaden. Thus, like the
previous case study, the OAU has not been able to play an
effective and decisive mediatory role in the Ethiopia-
Somalia dispute.

The Tanzania-Uganda Conflict

The Tanzania-Uganda conflict has some of the
ingredients of all three generic conflict-types that we
have identified in Africa. Initially, it was a challenge
to the integrity of the Idi Amin regime in Uganda and
also a personality conflict between Amin and President
Julius Nyerere to Tanzania. In late 1978, however, the
conflict was enlarged to become a challenge to the
territorial integrity of the Tanzanian State.
The initial conflict between Tanzania and Uganda
arose out of the overthrow of President Milton Obote of
Uganda by General (subsequently President) Idi Amin in
January 1971. Obote took refuge with a number of his
followers in neighbouring Tanzania. Nyerere refused to
accord legitimacy to the Amin government in Uganda for
two closely related reasons.   First, Obote was not only
a personal friend of Nyerere, he was also a seeming
socialist like the Tanzanian Head of State. And second,
Nyerere had a trenchant dislike for military regimes.
In September 1972 Amin alleged that his country was
invaded by pro-Obote guerrillas. He accused Tanzania
and Israel of providing assistance to these invaders.
Notwithstanding the fact that Tanzania did not reply to
the charges, Amin ordered his air force to bomb border
villages in Tanzania in retaliation for the alleged
invasion.[15]
Although Uganda did protest against the invasion to
the United Nations and the OAU, it was the latter in the
person of its Secretary-General that initiated the
process that led to a truce. However, once the peace
process had begun, the role of the OAU became peripheral.
The key mode of conflict control was a presidential
mediation pattern initiated by Somali President. Siad
Barre. Barre produced an agreement that led to the
initial containment of the conflict. Among other things,
Tanzania and Uganda agreed to withdraw their troops from
each other's territory and not to permit each other's
land to be used as a base for subversion. Somalia

supplied the observers to oversee the withdrawals.

The agreement that followed the Tanzania-Uganda conflict papered over the differences between the two antagonistic states. Uganda had wanted Obote handed over to it or, at a minimum, expelled from Tanzania. Tanzania, on the other hand, would like Obote to stay in the country. The issue of Obote and his loyalists in Tanzania stood at the root of the subsequent conflicts between Nyerere and Amin.

At the tenth anniversary summit meeting of the OAU in 1973, Amin and Nyerere met and the convivial ambience which such gatherings tend to create led to a super-ficial rapprochement between themselves and their two countries. But in October 1978, the Tanzania-Uganda conflict, which had been reduced since 1973 to occasional border clashes and sabre rattling by Uganda, broke out into open warfare. Uganda invaded Tanzania and declared that it was extending its border with that state to the Kagera river thus seizing 710 square miles of Tanzanian territory. In late November 1978, Tanzania launched a retaliatory raid against Uganda, although it too denied such an attack. Led and assisted by Tanzanian regular troops, anti-Amin groups in Tanzania overthrew the regime of Idi Amin in April 1979.

At the outbreak of the 1978 conflict, many African countries through their Heads of State appealed for a cessation of armed confrontation. Some leading African states such as Nigeria, tried presidential mediation. Idi Amin, while not discouraging such mediation by African Heads of State, appealed to the UN, the OAU, and the Arab League for assistance. However, African leaders dis-couraged him from taking the conflict to the UN thereby reaffirming the African position that inter-African con-flicts should be settled within the African forum, i.e., "try OAU first."[16]

The involvement of the OAU in the Tanzania-Uganda conflict in 1978 was direct and it came through its Chairman for 1978-79, President Numeiry of Sudan. Numeiry visited Tanzania and Uganda in December in his attempt to secure ceasation of hostilities. In November his special envoy, Phillip Obang and the OAU's Assistant Secretary-General for Political Affairs, Peter Onu, visited Uganda. This was in addition to the visit of the OAU Secretary-General, Edem Kodjo, to the warring coun-tries. These visits did not produce the desired cease-fire. The stumbling block which the OAU could not remove, because of the limitation of its Charter and its own in-adequate resources, were the pre-conditions set by Tanzania and Uganda for such a ceasefire.

On the one hand, Tanzania demanded the condemnation of Uganda for the October invasion which violated the OAU Charter. Uganda, on the other hand, requested, in early November, that the OAU should guarantee that Tanzania would never again invade it. The OAU could not fulfil

either of these separate preconditions. The reasons for
this were first, that as a collegiate, non-hierarchical,
and relatively egalitarian organisation, the OAU Charter
does not contain any sanctions mechanism. And second,
it did and does not have the resources to constrain
Tanzania and from assisting pro-Obote guerrillas.

Although Edem Kodjo had said in November 1978 that an
OAU committee would be established to resolve the border
dispute, such a committee did not come into fruition
until February 1979 when a nine-nation mediation committee
met in Nairobi. The members of the Committee were Central
African Republic, Gabon, Gambia, Madagascar, Nigeria,
Togo, Tunisia, Zaire, and Zambia.[17] But this Committee
was not able to arrange a ceasefire much less resolve the
conflict because of the Tanzanian delegation's insistence
on its earlier demand that Uganda should be condemned as
the aggressor. The meeting became deadlocked and the
Tanzanian representative withdrew.

The OAU Committee adjourned indefinitely four days
after the talks started and after it had resolved to
send missions to Uganda and Tanzania. The indefinite
adjournment was an unmistakable admission of failure by
the OAU. And its mission to Tanzania under Nigeria's
Foreign Minister could not convince Nyerere to drop his
pre-conditions so that the peace talks could continue.
Yet by the time the Heads of State and Governments
assembled in Monrovia in June 1979 for the 16th annual
summit meeting, the pro-Obote forces backed by Tanzanian
regulars, had succeeded in ousting Idi Amin from power.

Nevertheless, the 1979 OAU Assembly was seized with
the echoes of the Tanzania-Uganda conflict. The Nigerian
Head of State, General Olusegun Obasanjo, and the Sudanese
President, Jafar Numeiry, separately criticised Nyerere
for his alleged violation of the OAU Charter; he was
accused of invading Uganda and of overthrowing Idi Amin.

The majority of African leaders in the Assembly did
not share the views of Nigeria and Sudan, however. In
fact, many of them were silently happy at the termination
of a regime whose existence had become an embarrassment
to them because of its gross and widely publicised violation of human rights. President Toure of Guinea correct-
ly reflected this view when he told the Assembly that "if
Tanzania was guilty the OAU was also guilty; it had
failed to do anything about the enormities committed by
Amin and it had also failed to react effectively to the
earlier incursions by both sides."[18] The Assembly did not
pass any resolution to condemn Tanzania. Instead it
accepted and thus legitimised the regime assisted into
office in Uganda by Tanzania.

The Conflict in Western Sahara

The dispute over Western Sahara falls under the

category of threats to the integrity of the state. In this case, the state that is threatened is as yet only a potential one--Western Sahara. Morocco and Mauritania (before the latter pulled out of the territory in August 1979) had been trying to prevent the emergence of such an independent Sahara Arab Democratic Republic (SADR) in Western Sahara. Thus, the dispute is unique in the history of inter-African conflicts. Besides, it involves a threat to two fundamental principles of the OAU: the principle of the eradication of colonialism from the continent and the principle of self-determination for peoples in colonial territories. What, however, complicated the matter have been the claims of Morocco and Mauritania to the territory, which was originally administered by Spain. Mauritania and Morocco have both based their claims to the territory on historical and ethnic grounds; but both these claims are difficult to defend in the context of contemporary African international relations.

The OAU was seized with the Western Sahara problem in 1972 when the Council of Ministers expressed solidarity with the people of the territory in their struggle for independence from Spain. At the same time, the Ministers called on Spain to create what they called a free and democratic atmosphere which would allow the Saharans to exercise their right of self-determination. This basic principle of self-determination was again reaffirmed by the Ministers at their meetings in 1973 and 1974, respectively. From thenceforth, the Saharan problem featured regularly at the summits of the OAU and at the Ministerial Council meetings.

However, with the withdrawal of Spain from the territory in 1975, the OAU became directly enmeshed in the conflict. In fact, the dispute became a serious issue of concern to the OAU in June 1976 at the Council of Ministers meeting in Mauritius. There, the Republic of Benin tabled a pro-POLISARIO resolution which sought to enhance the status of the Saharois liberation movement. The resolution particularly called on the OAU to give its "unconditional support for the just struggle of the Saharan people for the recovery of their national rights."[19]

The AHG, notwithstanding this pro-Polisario resolution and the prevailing mood of the Ministers, was constrained to shelve the Saharan issue completely. The main reason for this action was pressure from both Morocco and Mauritania, which threatened to leave the Organisation if the matter was taken up by the summit. The Assembly decided instead to call another extraordinary summit which would be devoted entirely to Western Sahara. This was scheduled to take place in Lusaka on 25 April 1977. However, the summit could not be held because several African countries had "too many commitments to keep the scheduled date" and also because

Morocco "is now refusing to take part in OAU activities."[20]

At the 14th summit in Libreville, Gagon, in 1977, the African Heads of State again resolved to shelve any discussion of the Western Sahara conflict. They decided instead to call another emergency session of the Assembly which would be devoted entirely to Western Sahara. This was scheduled to take place in Lusaka, Zambia, from 5 to 10 October 1977. But for the second consecutive time an extraordinary meeting could not be held. The reason this time was what President Kaunda described as threats of aggression by the illegal regime of Ian Smith, although he was prepared to host the summit in January 1978.[21]

This further failure to hold the extraordinary session seemed to have given Morocco and Mauritania the upper hand in the dispute. At the same time, it exposed the inadequacy of the OAU's conflict control mechanism. As the Organisation did not have mandatory powers to force the recalcitrant Moroccans and Mauritanians to the conference table, there was a complete stalemate.

At the Khartoum summit in the Sudan, the OAU abandoned Conference Diplomacy for one of the more "practicable" and popular techniques of conflict control--the setting up of an ad hoc Committee of Heads of State. The Khartoum meeting thus called for the creation of a "Committee of Wisemen" under the chair of Sudanese President Numeiry to examine all the issues involved in the conflict and to report back to the Heads of State at the Monrovia summit in 1979.[22] This Committee met for two days in late 1978 in Khartoum. It detailed the Nigerian Head of State, Obasanjo, and Mali's President, Musa Traore, to contact all parties to the Western Sahara dispute. The main objective of this exercise, it seemed, was to secure a peaceful settlement of the conflict by the time the OAU held its next summit in Monrovia, Liberia, in June 1979. But the conflict remained unresolved. In fact, the Monrovia summit provided the first glimpse of the growing pro-Polisario feeling among African states. Following Mauritania's withdrawal from the Sahara, there was emerging a broad consensus that Morocco's policy towards the territory was not in conformity with the Organisation's anti-colonial policy. This feeling began to propel a large number of African states towards recognition of the SADR.

This growing anti-Morocco mood was apparent in the deliberations of the Council of Ministers which met in Monrovia in June 1979. The Ministers' resolution, among other things, called for (1) self-determination for the people of Western Sahara; (2) a referendum in the territory to be supervised jointly by the UN and the OAU; and (3) a complete ceasefire between Morocco and Polisario. As might be expected, the Moroccan delegates at the Ministerial Council meeting launched a strong attack and protest against what they saw as anti-Moroccan resolution by their colleagues. M. Boucetta, the Moroccan Foreign

Minister, accused the Ministers of "sabotage orchestrated by Algeria and its clients."[23]

The strong Moroccan protests notwithstanding, the AHG went ahead to consider the Western Saharan issue. In part, this move was influenced by the conciliatory attitude of the delegation of the new military regime in Mauritania towards Polisario. Nevertheless, the Heads of State did not come up with a radical solution to the problem. Indeed, they endorsed a Ministerial Council resolution which called for a meeting of the Ad Hoc Committee of "Wisemen" before the end of 1979, to work out the modalities of and supervise a referendum in the territory. Of course, none of the above recommendations were subsequently implemented by either the OAU or the parties to the conflict.

When the Ad Hoc Committee met in Monrovia in early December 1979 King Hassan was conspicuously absent. Nevertheless, he repeated his threat that his country would pull out of the OAU if the Saharan Arab Republic was recognised. At the end of the meeting, the Committee came up with a number of recommendations which again were a rehearsal of past OAU positions: (1) Morocco should pull out of the disputed territory; and (2) a call for a ceasefire which would be monitored by a UN peacekeeping force. The Moroccan King described the Committee's resolution as "tarnished with illegality and as such cannot have any obligatory character for us."[24] Legally, he was right. And politically, he did not heed the resolution, but there was little that the Committee or the OAU itself could do in the face of such defiance. The best the African Organisation could hope for in the circumstances was that the rising cost of the war would eventually force the authorities in Rabat to give up the territory.

In spite of its ineffectiveness, the AHG was still bent on calling the Moroccan King's bluff. The debate on Western Sahara at the 1980 summit in Freetown, Sierra Leone was very heated. President Samora Machel of Mozambique opened the discussion with a caustic attack on Morocco. He described the King's policy over the territory as colonialist and imperialist and called for the recognition of Polisario and the admission of the SADR into the OAU.

However, the conflict was never resolved at the Freetown summit. In fact, the issue proved to be the most divisive at the meeting and threatened the integrity of the OAU. It was obvious that many African states did not want to jeopardise the survival of the Organisation over the Saharan conflict. Thus at the end of the debate, during which the Moroccan delegation threatened to walk out together with its supporters--Egypt, Ivory Coast, Senegal, Sudan, and Tunisia--the call for the SADR's admission was supported by only twenty-six states. This was a simple majority in favour of Polisario, but the

Organisation did not formally admit the SADR. One of the reasons for this was fear that such an action would have led to the disintegration of the OAU. The other reason was Morocco's argument that admission would in fact require a two-thirds majority since the SADR was not a sovereign state. It was a complete stalemate; conference diplomacy could not resolve the Western Saharan conflict. Thus the AHG was constrained to resort to the less open technique of conflict resolution. The Ad Hoc Committee on Western Sahara was implored to "meet in the next three months" to find a permanent solution to the conflict.

The Committee met in Freetown from 9 to 12 September 1980 under the Chairmanship of Siaka Stevens of Sierra Leone. For the first time the representatives of Polisario and those of Morocco met face to face. This raised hopes that at long last peace would come to Western Sahara or at least that some interim solution would be found. However, the Committee did not make the expected breakthrough. Its recommendations merely regurgitated earlier OAU positions: a ceasefire and a referendum, to be implemented by December 1980. Again, there was in the end no ceasefire and no referendum in the territory, so the war continued.

Several factors were responsible for the apparent paralysis of the OAU's conflict resolution machinery over Western Sahara. First, is the lack of material resources with which it could embark upon an effective mediatory role. For instance, unlike the UN, the OAU is not in a position either to raise or finance a peacekeeping force to separate combatants in the battle field. The Chad debacle had already demonstrated this. Second, there is the patent absence in the Charter, as noted earlier in Chapter 1, of mandatory sanctions that could compel the disputants to reach a peaceful settlement. Third, the conflict had generated too much emotion in Morocco. The dispute is the only issue on which there is broad agreement between the right and the left of the Moroccan political spectrum. For the King, therefore, what was at stake was not only territory but also his throne and his prestige, both of which he might lose either by precipitate withdrawal or by agreeing to a solution that guarantees the self-determination of the Saharans.

Despite these obstacles, there was still some hope that the OAU would settle the conflict at the 18th summit in Nairobi. After the simple majority for SADR in Freetown it was, for instance, expected that the OAU would admit the territory as a member of Nairobi. Moreover, King Hassan was himself present at Nairobi at the head of his country's delegation. It was his first appearance at an OAU summit for several years. Besides, his speech to the AHG was rather conciliatory: "We have decided," he said, "to envisage a procedure of controlled referendum, the terms of which simultaneously respect the objectives of the latest Ad Hoc Committee of wise men and

the conviction Morocco has of its legitimate rights."[25]
The King's speech was ambiguous though. First, it is not
clear what he meant by Morocco's "legitimate rights."
Did it mean his country's historic claim over the territory as part of the greater Moroccan Kingdom? Second,
two days before the Nairobi speech, Hassan had declared
emphatically that his country "would not give up a grain
of sand" in the Western Sahara. It was doubtful, therefore, whether the King was genuinely interested in making
lasting concessions either to the AHG or to Polisario.
Finally, it is possible that the Nairobi speech was meant
to stem any drift towards recognition of Polisario and
the ultimate admission of SADR.

Nevertheless, the AHG welcomed the speech and the
King's open "commitment" to the idea of a referendum.
But again, the summit was anti-climactic. Resolution 103
(XVIII) merely called *inter alia* for: (1) an immediate
ceasefire; (2) the establishment of an "Implementation
Committee to be composed of Guinea, Kenya, Mali, Nigeria,
Sierra Leone, Sudan and Tanzania to ensure, with the
cooperation of the concerned parties and with full power,
the implementation of the recommendations of the Ad Hoc
Committee"; (3) the Implementation Committee to "meet
before the end of August 1981 and, in collaboration with
the parties in the conflict, to work out the modalities
and all other details relevant to the implementation of
the ceasefire and the conduct and administration of the
referendum";and finally (4) the Implementation Committee
to "take all necessary measures to guarantee the exercise
by the people of Western Sahara of self-determination
through a general and free referendum."[26]

The Implementation Committee met for the first time
in Nairobi from 24 to 26 August 1981 and took evidence
from Polisario and Morocco separately. At the end of the
two-day meeting, the Committee drew a detailed peace
package which provided for (1) a voter's list of all
Saharaouis from 18 years and above to be based on the
1974 census conducted by Spain; (2) "voting by secret
ballot on the basis of one person one vote"; and (3) the
Saharaouis to have the choice either for "complete
independence or integration with Morocco."[27] The
referendum was to be conducted under the auspices of the
UN and the Implementation Committee. Besides, the
recommendations provided that an "interim Administration
to be supported by civilian, military and police
components," to be set up in the territory. It also
provided for a ceasefire and a peacekeeping force to be
financed and assembled jointly by the OAU and the UN.
Finally, the result of the referendum was to be announced
to the Implementation Committee.[28]

Despite their detailed nature, the Nairobi peace
proposals did not lead to a resolution of the conflict
in Western Sahara. This was because both Morocco and
Polisario gave different interpretations of the Plan's

contents reflecting their particularistic interests. For
the Moroccans, the Implementation Committee's peace plan
meant (1) that there would be no face-to-face negotiations
with Polisario; (2) that its administrative and military
structures/presence in the territory would be maintained
during the referendum; and finally (3) that all Saharaouis
eligible to vote at the referendum would not only provide
proof of their identity but must also come under Moroccan
suzerainty.[29] For their part, Polisario spokesman
Mohammed Abdelaziz reasoned that the Implementation
Committee's decision meant among other things that there
should be direct negotiations with Morocco "in order to
launch a real, just and durable peace in accordance with
the inalienable rights of the Saharaouis."[30] With such
differing interpretations to the OAU peace plan by the
combatants, there was an obvious stalemate in the peace
process. Neither the OAU nor its Implementation Committee
possessed effective sanctions to compel either Morocco
or Polisario to modify their positions.

Three months after the Implementation Committee's
deadlocked peace plan, a meeting of the UN General
Assembly passed a very crucial resolution on the Western
Saharan conflict on 24 November 1981. For the first
time, the parties to the conflict were named: Morocco and
Polisario. The UN resolution urged both sides to "enter
into negotiations with a view to establishing an immediate
ceasefire" and to conclude a "peace agreement permitting
the fair conduct of a general, free and regular referendum
on self-determination in Western Sahara."[31]

It was under these circumstances that the Implementation Committee met again in early February 1982 in
Nairobi to try and break the stalemate. The meeting was
preceded by a preparatory meeting of the Ministers of the
seven member state from February 6 to 7. The Ministers
adopted three key resolutions on (1) a ceasefire;
(2) a referendum; and (3) what they called a UN role in
the peace plan for the Sahara. This last point is rather
interesting because, although the world body had always
been associated with the peace plan for the territory,
the Ministers deviated from precedent and recommended that
any peacekeeping in the territory would be the "complete
responsibility of the UN because of the exhorbitant cost
of the operation."[32] The reasons for their decision were
obvious. Since the OAU could not finance the peace-
keeping force in Chad, the Ministers were wary about its
ability to raise another force for Western Sahara.

The Implementation Committee met on 8 February but
ran into trouble soon after the meeting started. When
the Moroccan representative, Boucetta, was interviewed
by the Committee, he did not mince his words about his
country's position on the Saharan conflict. "To recognise the Polisario and SADR," he said, "you know our
position perfectly. Mr. Chairman . . . I must say here
solemnly that Polisario is not recognised as a liberation

movement. . . . Morocco intends to negotiate with
Algeria and Mauritania."[33] After the interview, the
Implementation Committee was left in no doubt that Morocco
was about to torpedo the peace process. To avoid another
stalemate, the Committee sidestepped the UN resolution
which identified the parties to the conflict to be
Morocco and Polisario and merely called on the "parties
to the conflict" as well as "interested parties" to find
a peaceful solution. Since it refrained from naming
Morocco and Polisario as the parties to the conflict,
the Committee could not compel them to adhere to a ceasefire. Nevertheless, it imaginatively called on the
parties to the conflict to observe a "total ceasefire
which will take effect at a date to be fixed by the
Implementation Committee upon the advice of its chairman
after consultations with all the concerned parties."[34]
The "parties concerned" did not heed the ceasefire call.
Thus, the third attempt to settle the conflict in
Nairobi--called Nairobi III--like all previous attempts
failed to bring peace to Western Sahara.

After the collapse of Nairobi III the admission of
Polisario into the OAU seemed to be the only option left
to the pro-SADR faction within the Organisation to compel
Morocco to return to the peace table. Consequently, at
the 38th Ministerial Council meeting in February 1982--
to consider routinely the OAU budget--the SADR delegation
was allowed to take its seat between the Nigerian and
Rwandese delegations, thus signalling formal admission
into the OAU.

The decision took many members by surprise. The
Moroccan government, as would be expected, was most
shocked by the Secretary-General's action. Morocco,
which was represented at the meeting by its ambassador
to Sudan, called the admission of the SADR as a
"manoeuvre" by the Secretariat and the ambassador re-
emphasised his country's position that it "will not sit
at the same table as representatives of an imaginary
republic." He therefore walked out of the conference
room followed by pro-Morocco supporters. The OAU was
thrown into the most divisive crisis since its inception
in 1963.

From Rabat, King Hassan described the decision by the
Ministers to admit the SADR as a "juridical hold-up
(and an act) of banditry." He accused Secretary-General
Kodjo of "abuse of power and taking a measure which in no
way is among your attributes or prerogatives." The King
concluded his attack by saying that his country expected
the Secretary-General to "take all adequate or necessary
measures to restore the OAU legality and that the illegal
measure tending to recognise the SADR as a possible
member of the OAU be revoked."[35]

By the time the Ministerial Council meeting ended on
28 February, a total of nineteen states, including
Morocco, had walked out of the conference in protest

against the seating of the SADR delegation. Surprisingly,
though, Morocco did not tender its resignation from the
OAU as it had always threatened to do. Instead, the King
mounted a diplomatic campaign--backed secretly by the
West--to prevent the Tripoli summit from being held in
August 1982.

The 19th annual summit in Tripoli was not held for
lack of a quorum--the number of states needed to form a
quorum is thirty-four--but there were only thirty-two
states present in Tripoli. All the pro-Morocco states
stayed away. Besides, many conservative states did not go
to Tripoli. To their leaders, the thought of having
Gaddafy, whom they accused of engineering political dis-
turbances and coups, as Chairman for 1982-83 was unacceptable. Other reasons for the failure of the Tripoli
summit include the inability of the then Chairman,
President Arap Moi of Kenya, to attend. He was forced to
stay at home by an abortive attempt by his air force to
topple his government on 1 August 1982. Finally, the
Tripoli summit failed because the OAU, "which was founded
to achieve political freedom for Africa, has almost
reached its goal."[36] And according to this argument, the
Organisation should now turn its attention to "economic
development as the continent's new objective."[37] Whatever the reasons for the failure of the Tripoli summits,
one thing is clear; this is that the OAU has throughout
its twenty years' existence failed hopelessly to find
peaceful solutions to inter-African conflicts, as our
case studies have adequately demonstrated.

CONCLUSIONS

We have argued in this chapter that despite the
development of elaborate conflict control mechanisms,
the OAU was not able to find lasting solutions to the
disputes studied. The efforts of the Organisation in the
four case studies demonstrated very convincingly its
ineffectiveness in managing inter-African disputes. One
reason for the OAU's powerlessness is its low resource
base and Charter limitations. Although all four conflicts were referred to the Organisation by one of the
parties to the disputes, it could not, as in the Tanzania-
Uganda conflict for instance, persuade the combatants
to agree to a ceasefire nor could it, as in the Western
Saharan case, compel Morocco to honour its resolutions
urging a referendum and ceasefire in the Sahara. Similarly, the OAU at the time of writing has not done anything
about the renewed fightings in the Ogaden. Finally, as
noted in the Chad case, the OAU peace force did not
succeed in imposing a solution on the warring factions
in Chad.

The intervention of the OAU in two of the four
conflicts through Ad Hoc Committees was limited to

maintaining a presence and legitimising "Presidential
Mediation" initiated by individual African states such as
Nigeria in the Chad conflict and Somalia in the Tanzania-
Uganda dispute. Notwithstanding the OAU's ineffectiveness
in these conflicts, the Organisation is functional for
African states in terms of helping to insulate their conflicts from super-power rivalry; or at least in terms of
giving them more control over such extra-continental
intervention.

Many suggestions have been advanced to solve the
OAU's weakness in the realm of conflict control. It is
argued that sanctions should be written into the Charter
while others have put forward the idea of a mini "Security
Council" for the OAU. These suggestions are laudable
but miss the point for a variety of reasons. First, it
is doubtful that any effective machinery could be put
into place to enforce such sanctions. Second, African
regimes are notorious for their fickleness towards constitutions. On the basis of this reasoning, it is doubtful whether OAU states would faithfully implement any
sanctions that an amended OAU Charter might contain. The
clandestine trade links which some African states have
with the pariah regime in South Africa despite routine
OAU resolutions against such links is a sobering example.
And third, a final reason why a mere change in the Charter
may not make for an effective conflict control by the OAU
relates to resources. It has been clearly demonstrated
that the Organisation does not have resources to enforce
its resolutions. Hence, the provision of sanctions in the
Charter without the requisite institutions and resources
to enforce them will make the OAU a laughing stock.

NOTES

1. For details, see Chapter 1.
2. Oran Young, Intermediaries in World Conflict:
Third Parties in International Politics (Princeton:
Princeton University Press, 1967).
3. OAU: Introduction to the Report of the Secretary-
General on the Activities of the Organisation (CM/1002/
XXXIV) (Addis Ababa, 1980) pp. 7-9.
4. For more details on the problems of an African
High Command, see Orobola Fasehun, "Nigeria and the Issue
of an African High Command: Towards a Regional and/or
Continental Defence System," in Afrika Spectrum 80/3,
pp. 309-317.
5. Keesings Contemporary Archives (KCA), 1 February
1981, pp. 30064-67.
6. West Africa (London), 6 July 1981, p. 1513.
7. West Africa, 2 November 1981, p. 2555.
8. West Africa, 23 November 1981, p. 2757. Togo did
not supply troops because Weddeye objected to the statement by Eyadema that a political solution was necessary.

Benin and Guinea could not finance their own troops in Chad.

9. West Africa, 21-28 December 1981, p. 3096.
10. West Africa, 30 November 1981, p. 2820.
11. West Africa, 5 July 1982, p. 1781. The fact that individual countries were to be responsible for the upkeep of their troops made it possible for them to turn to external sources for financial assistance. This in turn jeopardised the neutrality of the force.
12. West Africa, 5 July 1982, p. 1781.
13. For details on the deliberation of the meeting, see CM/Res4(11), 1964.
14. For details, see ibid.
15. ARB, September 1972, p. 2597.
16. On the concept of trying the OAU first, see Berham Andemicael, The OAU and the UN (New York: Africana, 1976).
17. ARB, February 1977, p. 1357.
18. West Africa, 30 July 1979, p. 4525.
19. See ARB, July 1976, p. 4081.
20. Africa Contemporary Record (ACR) Volume 1976/77 (London: Rex Collings), p. PC 23.
21. See "Western Sahara OAU Summit Postponed" West Africa, 19 September 1977, p. 11951.
22. The other members of the Committee are Guinea, Mali, Nigeria, and Tanzania. See ARB, July 1978, p. 4914, for more details on the resolution.
23. KCA, 21 September 1979, p. 2941.
24. Africa Diary (New Delhi) 15-21 April 1980, p. 9975.
25. Quoted by Celia Seggel Pangalis, "Conflict in Western Sahara" (New York: International Peace Academy, November 1981), p. 20.
26. AHG/Res. 103 (XVIII), pp. 1 and 2.
27. Pangalis, "Conflict in Western Sahara," p. 20.
28. OAU Implementation Committee on Western Sahara meeting in its first session in Nairobi from 24-26 August 1981, pursuant to resolution AHG/Resl 03 (XVIII), pp. 1-3.
29. Africa News, 31 August 1981, quoted in Pangalis, "Conflict in Western Sahara," p. 30, footnote 44.
30. Weekly Review, 28 August 1981.
31. ARB, February 1982, p. 6356.
32. ARB, 1-28 February, p. 6356. The members of the Implementation Committee are Guinea, Kenya, Mali, Nigeria, Sierra Leone, Sudan and Tanzania.
33. Ibid, p. 6356. He later told reporters that "there will be no withdrawal of Moroccan troops from the territory. There will be no retreat of Moroccan administration from the territory."
34. Ibid., p. 6357.
35. Ibid.
36. President Didier Ratsiraca of the Malagasy Republic in Daily Sketch (Ibadan), 31 August 1982.
37. Ibid.

4
OAU and Economic Development in Africa

The African leaders that assembled in Addis Ababa in 1963 appreciated the enormous economic and social problems which the continent faced. They inherited an economy which was ill suited to the needs of their newly independent states. The economy had been designed mainly to serve the interests of the erstwhile colonial powers. It was tailored to produce primary agricultural and mineral resources which were exported to the colonial metropoles. In return, the African states imported finished products. The terms of trade for Africa's exports were poor. Consequently, there was a trade imbalance heavily in favour of the developed countries. The infrastructures inherited were poor and they served to enhance the colonial economy. Roads, railways, ports, and telecommunications were not only scarce but they also hardly penetrated the hinterland. Roads and railways run from agricultural and mineral producing centres to the coast, while telecommunication links were between metropolitan and African capital cities. Besides, there was a high population growth rate, which adversely affected economic growth.

The assembled African leaders were also cognisant of the fact that the security and stability of their countries individually and the continent as a whole would in the long run be inseparable from their ability to fulfil the rising economic expectations of their peoples. Economic prosperity of the continent would not only promote national concord, but would also enhance the political stability of the newly independent states. This would help them face the arduous task of nation building.

Not surprisingly, one of the purposes of the OAU was to "coordinate and intensify cooperation and efforts to achieve a better life for the peoples of Africa." Thus at the founding conference of May 1963, the African heads of state decided to create an Economic and Social Commission as one of the Specialised Commissions of their new organisation. To that effect, they set up a

Preparatory Economic Committee to study the whole question
of economic cooperation and development among members,
in collaboration with the Economic Commission for Africa
(ECA). Specifically, the Committee was charged with the
task of studying:

1. the possibility of establishing a free trade
   area between the various African countries;
2. the establishment of a common external tariff
   to protect the emergent industries and the
   setting-up of a raw materials price stabili-
   zation fund;
3. the restructuring of international trade;
4. the means of developing trade among African
   countries by the organization and participation in African trade fairs and exhibitions
   and by granting of transport and transit
   facilities;
5. the coordination of means of transport and the
   establishment of road, air, and maritime
   companies;
6. the establishment of an African Payments and
   Clearing Union;
7. the establishment of a Pan-African Monetary
   zone; and
8. the ways and means of effecting the harmoniza-
   tion of existing and future national development plans.[1]

OAU's PERFORMANCE: DECADE OF NEGLECT, 1963-1973

The Economic and Social Commission held only two
meetings. The first meeting was held in Niamey in
December 1963 and the second one in Cairo in January 1965.
The latter, significantly enough, was devoted almost
exclusively to social development matters. After the
Cairo meeting, it became difficult for the Commission to
meet. One of the reasons for this was the inability to
form a quorum at planned meetings of the Commission.
This situation, according to Diallo Telli, the first OAU
Secretary-General, was due to the non-response to African
countries to invitation for the Commission's meetings.[2]
It would appear that African states were, at the time,
too pre-occupied with their individual development
strategies and therefore neglected the Commission.
Nonetheless, African leaders continued to pay lip
service to their collective determination to promote the
continent's economic development. At the Algiers summit
in 1968, they once again reaffirmed their allegiance to
the Charter provisions with regard to economic matters.
Specifically, they pointed to the urgent need to
accelerate the pace of the continent's economic development as constituting an essential condition for the

fulfilment of the aspirations of the African peoples. Two years later, they again recommended the intensification of regional cooperation, the harmonisation and coordination of legislation and customs procedures as well as monetary cooperation.

OAU's poor performance in the sphere of economic development in this period can easily be explained by a number of factors. First, the annual summits which became a *de facto* substitute for the Economic Commission were not the proper fora for detailed and purposeful examination of Africa's development strategies. The summits were not equipped with the technical expertise required on such matters. Not surprisingly, the African leaders limited themselves to the passing of pious resolutions which had no practical effect.

Second, the OAU was preoccupied with political issues to the utter neglect of economic development issues. The leaders were more concerned with political stability and territorial integrity of their states--issues that were of more immediate concern to them throughout the Organisation's first decade. And third, the disparate political and ideological systems of member countries made it very difficult for the Organisation to perform. As Nzo Ekangaki lamented in his review of the Organisation's performance in its first decade: "OAU's economic and social role was relegated to the background as priority was not given to economic issues at that juncture."[3]

Between 1963 and 1973, the OAU tacitly abdicated all responsibility in the economic development sphere to the better organised and technically more capable ECA. A number of important conferences were organised by the ECA in this period. One of such meetings was the ministerial Conference on Trade Development and Monetary Problems held in Abidjan, Ivory Coast, in May 1973. It brought together the Ministers of Trade and Development of member states of the OAU. The most salient outcome of the four-day meeting was a draft declaration on "Cooperation, Development and Economic Independence." The document was subsequently adopted by the tenth Anniversary Summit in Addis Ababa as "The African Declaration on Cooperation, Development and Economic Independence," the Addis Ababa Declaration for short.[4]

The Addis Ababa Declaration was broad in scope and covers areas such as energy, natural resources, agriculture, transport and telecommunications, industrialisation, monetary and financial matters, the environment, development finance, and international trade. Like earlier resolutions, the Addis Ababa Declaration is notable for its declaratory and recommendatory nature. Besides, it seems to raise more questions than it actually attempted to answer. On the issue of Africa's natural resources, for instance, the Declaration urged African States individually and collectively to "undertake a systematic survey of all Africa's resources, with a view to their

rational utilisation and exploitation." African countries
were also urged in the Declaration to "defend vigorously,
continually and jointly (their) inalienable sovereign
rights and control over their natural resources."
Furthermore, they were to "promote the modernisation
of African agriculture through the introduction of modern
and advanced techniques and through the exchange among
African countries of experience and methods in the field
of production, distribution and storage." Finally, the
Declaration called on member countries of the OAU to
"take adequate measures to ensure rational industrialisation, within the context of sub-regional and continental
economic entities."

Despite the good intentions behind the Declaration
and the pious exhortations, it seemed that the African
leaders overestimated their own capabilities individually
and collectively. It is doubtful, for instance, whether
the injunction on natural resources could be implemented.
First, there is no way in which African states could have
defended "vigorously, continually and jointly (their) inalienable sovereign rights and control over their natural
resources" given the paucity of even the basic technology
needed to map out the continent's available natural
resources, not to talk of their economic exploitation.
And second, the sheer poverty of most African countries
makes such ventures well beyond their capabilities.
These twin problems have in the past, as it is at present,
continued to give rise to unequal treaties and con-
cession agreements between African countries on the one
hand, and foreign multinational corporations on the other,
which are in possession of valuable technology as well as
badly needed funds to undertake surveys, exploration, and
exploitation of the continent's natural resources.

THE DECADE OF ENFORCED INVOLVEMENT,
STRATEGIES, AND PLANS, 1973-1983

By 1973 a confluence of unrelated external factors
compelled the OAU to pay more attention to economic
issues. Notable among these factors were the impending
negotiations with the European Economic Community (EEC)
and the oil crisis of October that year. At the Tenth
Anniversary Summit in Addis Ababa, the general mood of
the African leaders was that they needed to coordinate
urgently their economic policies in order to release the
continent from its perpetual subservience to external
powers. They had in mind, in particular, the previous
agreement between the EEC and the so-called Associated
African States, which they viewed as grossly unequal and
detrimental to African interests. They believed that the
shortcomings in the agreement could have been avoided if
African States had negotiated as a group.

Accordingly, they decided to adopt three broad strategies with regard to future negotiations between Africa and the foreign powers:

(i)   Take necessary precautions in their negotiations with developed countries either individually or collectively to ensure that they do not become subjected to any foreign economic power

(ii)   Concert and organise their action in advance of all negotiations with the developed countries in order to assess all the implications which the proposed agreements might have on the future of their economic independence (regarded inviolable principle);

(iii)   Act collectively in multilateral trade negotiations in order to safeguard a number of objectives among which are:

    (i)   the adoption of effective concerted measures to put a definite end to the constant deterioration in the terms of trade of African countries;

    (ii)   the abolition by the developed countries of all tariff and non-tariff barriers and restrictive trade practices which those countries have hitherto placed in the way of the penetration of their markets by products from the African countries;

    (iii)   non-reciprocity in trade and tariff concessions accorded to African countries by the developed countries; and

    (iv)   the adoption and effective implementation by all the developed countries of the generalized system of preferences, the suppression of all escape clauses, the extension of the system to cover all African exports and its adoption by all countries that have not yet done so.[5]

The Heads of State mandated trade and economic development ministers to meet and work out the details of the African strategy. The Ministers met in Dar-es-Salaam in October 1973. Among other things, they agreed to adopt a common position for the forthcoming negotiations with the EEC for a new agreement that would cover both the then associated Francophone African states and the others.

This was a very significant achievement for the OAU given the diversity of interests among its members and in

particular, the opposition from the associated states.
This is not to suggest, however, that such unity of
position in negotiations cannot produce a neo-colonial
arrangement.

The Oil Crisis

The institution of an Arab oil embargo against
Western supporters of Israel, their production cut-back
and their unilateral and unprecedented oil price increases in October 1973, as part of the Arab war effort
against Israel, had a disruptive effect on the world
economy. Although these measures were not primarily
directed against African states, they nonetheless had a
devastating impact on the already frail economies of
African countries.

The adverse impact of the "oil weapon" on Africa was
predictable. Except for the four African producers of
oil, African states were net importers of oil. More significantly, they were almost totally dependent on Arab
oil. Thus, within a few weeks of the Arab oil measures,
most African countries were experiencing serious oil
shortages. They adopted stringent measures such as steep
price increases, rationing, and outright prohibition of
driving of private cars and selling of petroleum during
certain days of the week. For example, Ghana increased
the prices of premium grade oil by 60% in February 1974
and the prices of kerosene almost doubled despite the
government subsidy of over $25 million a year.[6]

By early 1975, Ghana's import bill for mineral fuels
alone was running at over four times the 1974 level. In
Senegal, petrol prices rose by between 26-29%, and in Chad,
by 100% at the beginning of February 1974. In Sierra
Leone the oil import bill jumped from Le 7.5 million in
1973 to Le 21 million in 1974. The pattern was the same
throughout the continent. According to an OAU study
in early 1974 it was estimated that eleven non-oil producing members with their own refineries would have to
pay $635 million for crude oil in 1974 compared with
$185 million in 1973, an increase of 242%. Moreover, this
was only if their crude oil consumption remained at the
1973 level and prices remained at $12 per barrel.[7]
Besides, African economies were interwoven with those of
the developed countries of the West, who were the major
target of the Arab oil weapon.

The combined effect of this on Africa was twofold.
First, the inflation and recession in the West consequent
upon the Arab oil measures led to a drastic fall in the
demand for Africa's export commodities. The commodity
boom which had begun in 1972 came to an abrupt end. This
situation was particularly bad for the African countries,
most of which depended on single export commodities for
their foreign exchange earnings. Second, African states

had to pay substantially higher prices for imports from Western countries. For example, there was a fourfold increase in the price of fertilizer between 1973 and 1975. In 1973, before the oil measures, a ton of "standard" fertilizer cost only £25. However, by 1974 it had risen to £66 reaching an all high price of £100 in 1975.[8] With only a small increase in the volume of total imports, the import bills of the African states rose phenomenally as the higher fuel prices worked their way through the economies of the developed countries.

The OAU's Response to the Oil Crisis

The oil crisis left African states with no option other than to try to seek drastic solutions to their deteriorating economic plight. For the first time, the OAU spontaneously and unanimously saw the need for collective action. An extraordinary ministerial council was convened in Addis Ababa in November 1973 to examine the impact of the oil crisis on member states. At the meeting, the ministers set up a committee of seven comprising Botswana, Cameroon, Ghana, Mali, Sudan, Tanzania, and Zaire--otherwise known as the Committee of Seven--to study the effects of higher prices and oil shortages on Africa and to find ways and means of alleviating the problems caused by the crisis. Significantly, all the members were non-oil-producers.

The committee held its first meeting in Addis Ababa in late December 1973. It considered a report submitted by the Secretary-General on the effect of the crisis on African economies. The committee was convinced that the problems caused by the crisis could only be resolved through a dialogue with the Arab oil states.[9] This perception stemmed from a number of factors. First, as we have already pointed out, the African countries depended almost entirely on Arab oil. Second, was the newly found solidarity between the Arab and African states following the support Africa gave the Arabs in the October 1973 Middle East War.[10] This made Africans believe that the Arabs would reciprocate their solidarity by ensuring the uninterrupted supply of oil to them and at concessionary rates. They therefore recommended a joint meeting of Arab Foreign Ministers and Ministers of Petroleum with the OAU committee.

The African Committee held a series of meetings with the Arabs between January 1974 and March 1977. However the Arabs rejected the major demands of the African countries: (1) oil at preferential rates; (2) zoning of African countries with each zone to be supplied oil by an Arab or a group of Arab countries; (3) direct agreement with Arab oil producers without intermediaries; (4) sale of oil at c.i.f. to African states; and (5) financial assistance in procuring tankers. This last demand would

have given African countries direct access to Arab oil and
would have made them less dependent on western oil multinational companies. A positive response by the Arabs to
the African demands would have also strengthened the OAU
as an organisation and the secretariat in particular.
This would have encouraged the individual OAU member
states to repose more confidence in the ability of the
organisation to effectively respond and protect their
collective interest. For the secretariat especially, there
was high hope that successful negotiations with the Arabs
would have transformed it from an administrative to a *de
facto* executive organ. This explains the great enthusiasm
exhibited by the secretariat initially in responding to
the OAU's mandate as well as the disappointment felt after
the failure of the protracted negotiations with the Arab
League.[11] For the African states, the failure to get
concessions from the Arab states meant there would be no
relief for their troubled economies. Their problems were
further compounded by the escalating oil prices.

The Era of Strategies and Plans, 1976-1980

The devastating effect of the oil crisis on African
economies brought home dramatically to African states the
need to pay more collective attention to economic issues.
The negotiations with the Arabs, in particular, demonstrated more than ever that for the continent to be in a
stronger bargaining position vis-a-vis other groups in
the international system, it must be economically strong
and politically united.

Not surprisingly, 1976 saw the first special session
of the ministerial council meeting of the OAU which was
devoted exclusively to economic issues. The four-day
meeting was held in Kinshasa in December. The council
recommended the creation of an African economic community
along the lines of the EEC over a 15-25-year period. Besides, the ministers decided that an integrated energy
policy should be worked out by the OAU Secretary-General
and the ECA Executive Secretary. They also recommended
that the ECA together with the OAU should set up an
economic data bank as soon as possible to facilitate the
realisation of the projected economic community. Finally,
on North-South Dialogue, the ministers reaffirmed the
Organisation's determination to "promote and strengthen
African cooperation for the good of the African peoples
in conformity with the aims in the OAU Charter and in the
Organisation's Declaration on Cooperation for Development
and Economic Independence."[12]

At the rhetorical level, the conference reaffirmed
Africa's determination to come to grips with its worsening economic plight. In reality, however, its main
recommendation, that of the creation of a common market
was in 1976 neither a new idea nor was it a realistic

proposition. Experience has shown that sub-regional economic integration in Africa, where this is possible, is fraught with political, ideological, economic and personality problems. A clear case in point is the now defunct Economic Community of East Africa. The Community brought together three contiguous former British colonies; Kenya, Tanzania, and Uganda. Nevertheless, the Community was bedeviled by a lot of problems such as the personality differences between General Idi Amin and President Nyerere on the one hand, and between the socialist and capitalist leaderships in Tanzania and Kenya respectively. Besides, the cooperating states adopted different economic strategies. The confluence of these factors made it impossible for the Community to continue to function and finally resulted in its collapse in 1977. In some other sub-regions of the continent it has been difficult to form a regional economic grouping. So far only the West African Economic Community, which groups about sixteen English- and French-speaking states, has made a small progress towards a common market. But it too is fraught with many problems. Besides, a common market with essentially capitalist orientation may further facilitate the exploitation of weaker states by the more powerful members. It is therefore illusory to think that an African economic community would be formed as the political, cultural, linguistic, ideological and economic complexities are bound to be magnified at the continental level.

Nonetheless, the ministerial conference was a significant landmark in the OAU's efforts to find lasting solutions to Africa's economic problems. It renewed interest in the idea of Pan-African economic integration. More significantly, it led to increased joint collaboration between the OAU and the ECA in trying to evolve permanent strategies with regard to the continent's future socio-economic development and industrialisation. Such cooperation is important given the superior financial and organisational resources available to the ECA.

OAU-ECA collaboration became very evident in February 1979, when both institutions co-sponsored a symposium on the Future Development Prospects of Africa Towards the Year 2000. The symposium was devoted to an analysis of the causes of Africa's economic backwardness with a view to mapping out strategies which could lead to its economic and industrial take-off by the turn of the century. The symposium discovered among other things that the present poor economic performance of African states was a result of a combination of factors: (1) the lopsided nature of the international economic system in favour of the industrialised countries at the expense of the developing countries especially those in Africa; and (2) misconceived and "erroneous strategies that have steered them towards ill-suited models of development that are geared neither to human needs nor to a basically endogenous development."[13]

In the light of the above analysis, the symposium agreed on the need for a complete break with past strategies of development. Henceforth, "the areas which depend on domestic policies, structural changes and systems of values must be given priority attention so that a new human-being-oriented African development policy can evolve." Under this new strategy priority should be given to the creation of a material and cultural environment that is conducive to self fulfilment and creative participation. Besides, there should be a rational use and exploitation of Africa's resources to ensure self-sufficiency in food and local processing of raw materials.

Finally, the symposium recommended a four-pronged "strategy for change", which would entail:

    (i)   a new pedagogy geared to African Unity;

    (ii)   the need for scientific, cultural and social values underlying a new approach to development;

  (iii) mastery over the technical and financial instruments that are vital to the new type of development; and

    (iv) a new approach towards international cooperation, with the emphasis on links between developing countries.[14]

The recommendations of the Monrovia symposium seem to be more realistic than their Kinshasa counterparts. For instance, although both meetings recommended the creation of an African common market, the Monrovia symposium called for a "progressive coordination and integration of the existing economic regions." For a start, the projected common market need not involve all spheres of inter-state economic activity. It may be applicable to specific commodities and products, such as meat, cereals, and so on. Second, "the extension of arrangement for the free movement of persons and goods on the African continent." In other words, African states should relax visa requirements for Africans and impose less restrictions on inter-African trade. And third, African leaders should mount an enlightenment campaign in their respective countries to gain support "for the idea of the unity of Africa so that it does not remain the preserve of political circles alone."[15]

The recommendations of the Monrovia symposium formed the basis of discussion at the sixteenth ordinary session of the AHG in Monrovia in 1979. For the first time, the summit devoted a considerable proportion of its time discussing economic issues. The summit adopted the Monrovia Declaration of Commitment of the Heads of State and Government of the Organisation of African Unity on Guidelines and Measures for National and Collective Self-Reliance in Social and Economic Development for the Establishment of a New International Economic Order (henceforth referred to as the Monrovia Declaration).

The Monrovia Declaration is significant in many respects. First, it goes further than the usual OAU resolutions in committing the Heads of State "individually and collectively on behalf of (their) governments and peoples" to redress the precarious economic plight of the continent. Second, and more significantly, it recorded their commitment "to give special attention to the discussion of economic issues at each session of our Assembly." To effect this, the AHG mandated the OAU Secretary-General in collaboration with the Executive Secretary of the ECA to draw up annually "specific programmes and measures for economic cooperation on sub-regional, regional and continental basis in Africa."[16]

The meeting also decided to hold a special economic summit in Lagos before the end of 1979. It instructed the OAU secretariat in collaboration with its ECA counterpart and ministers of member states responsible for economic development to prepare the ground for the proposed extraordinary session "so as to facilitate a fruitful discussion of the economic problems of Africa."

However, because of the transition from a military to civilian administration in Nigeria in October 1979, the proposed meeting could not be held until 28 April 1980. The deliberations of the meeting were heavily influenced by the gloomy economic situation prevalent on the continent. Although Africa is endowed with vast mineral and agricultural resources, it remains nonetheless the least developed continent. It contains over twenty of the thirty-one least developed countries of the world. The total Gross Domestic Product of all the countries in Africa was in 1980 only 2.7 percent of the total world product, and per capita income in Africa averaged only $166,[17] pitifully below the average for the less developed countries (LDCs) as a whole, which was in 1975 roughly estimated to be $300.[18]

Besides, the food situation in Africa had considerably deteriorated from what obtained at the time of independence. The Director-General of the Food and Agricultural Organisation of the UN (FAO) Edouard Saouma dramatically put it in his address to the summit: "Today, the average African has 10 percent less food than 10 years ago. Average dietary standards have fallen below essential requirements. Hunger and malnutrition afflict more and more Africans. Starvation faces millions of refugees."[19]

The prime consideration of the summit was, therefore, how to achieve a self-reliant, endogenous economic development. The summit consequently adopted a blueprint for Africa's economic development called the "Lagos Plan of Action for the Economic Development of Africa 1980-2000," hereafter referred to as the Lagos Plan of Action.

The Lagos Plan of Action

The Lagos Plan of Action mapped out four broad priority areas for Africa's development by the year 2000. These are food and agriculture, energy, trade and finance and the creation of an African common market.

Food and Agriculture. The issue of how to feed the continent's population both at present and in future is an important one. Africa has one of the world's highest population growth rates. It is expected that the population will rise from 406 million in 1980 to 827 million by the year 2000. The implication of the rapidly growing population is that the continent has become more and more dependent on external sources for food supplies and/or money (loans) to pay for such supplies.

Not surprisingly, therefore, the Lagos Plan stressed self-sufficiency in food production, as one of the main prerequisites for the continent's endogeneous development by the year 2000. According to the Plan, African leaders not only agreed to give priority to food production but also to channel a greatly increased volume of resources to agriculture. It further enjoins African leaders to initiate policies that would encourage small farmers and members of agricultural cooperatives to achieve higher levels of productivity. In order to reduce the ever growing African dependence on foreign sources, the leaders agreed to give special attention to the cultivation of cereals, such as millet, maize and sorghum, so as to replace wheat and barley. In addition, they also decided to undertake action aimed at securing a substantial reduction in food wastage, especially post-harvest losses.

Energy. The problems identified in this section of the Plan are more daunting and striking than those for food. At the moment, although Africa is richly endowed with energy generating raw resources, it has nevertheless relied on extra-Africa countries almost exclusively for its energy needs. African countries, including OPEC members, neither have the funds nor the technology to economically tap their natural resources. In fact, in most African countries, such resources are yet to be properly identified and mapped out. Consequently, the control which African states have over their strategic mineral resources is very tenuous indeed.

The Plan proposes to solve Africa's energy problems in two stages; viz, short-term, medium and long-term. Measures envisaged under the short-term arrangement include

1. stable and guaranteed supply of oil to African countries;
2. increased technical assistance to non-oil producing countries by African OPEC members in such areas as staff training and in prospecting for and exploiting oil deposits;

3. arrangements to enable oil importers to
   obtain their supplies at concessionary
   prices or compensation fund to partially
   offset their worsening balance of payments
   deficits caused by rising crude prices.

Under the medium and long-term strategy, the Plan
identifies four broad areas that require concentrated
collective attention. These are (1) enhancing the development and utilisation of the supply of fossil fuels;
(2) development of hydropower resources; (3) development
of new and renewable sources of energy; and (4) development and utilisation of nuclear energy.

Trade and Finance. Under trade and finance, the Plan
recommended, among other things, that "particular attention should be given to domestic trade and to improving
the conditions under which it is now taking place." As
for intra-African trade, the Plan recommended the reduction and eventual elimination of trade barriers and
also suggested mechanisms and measures which would
facilitate the development of trade among African countries. In that regard, the Lagos Plan urged the OAU, ECA
and other competent organisations to undertake studies on
the demand and supply of major intra-African trade
commodities.[20]

The Plan further recommended measures which could
lead to a diversification, both geographically and
structurally, of Africa's existing international trade
links. It is envisaged that by the end of the century,
Africa's share of trade in manufactures would have reached
the 25 percent target which has been set for developing
countries by the UN.

Besides trade, there are also elaborate provisions on
finance in the Plan. At the national level, African states
agreed to completely restructure their monetary and
financial institutions and to ensure a more judicious
management of limited financial resources. At the sub-
regional level, African states decided to integrate their
financial systems not later than the end of 1984. They
further agreed to establish clearing and payments arrangements which would form the nucleus of an African payments
union before 1990. At the continental level, they agreed
to strengthen the ADB to enable it to give more financial
assistance to member states and to create an African
monetary fund. Finally, it was agreed in Lagos that
African states should intensify their efforts towards the
establishment of a New International Economic Order (NIEO).

The Lagos Plan envisages the creation of an African
common market by the year 2000 in two phases. During the
first phase, 1980-85, the leaders pledged themselves to
create and strengthen sub-regional economic groupings
which will facilitate continental economic integration,
and to create similar institutions where they do not
already exist. It is hoped that the successful

implementation of the first phase would inexorably lead
to the second phase--the creation of the African common
market itself by the year 2000.

As a blueprint, the Lagos Plan of Action is undoubtedly a beautiful and an invaluable document. Indeed,
it has got all the ingredients that would in theory lead
to an endogenous African development. However, the Plan
is fraught with a lot of problems. One obvious shortcoming of the Plan is that it is merely recommendatory.
Thus member states can choose to ignore it. This perhaps
explains in part why nothing has been done since its
euphoric adoption in 1980. In spite of this major shortcoming, the Plan goes on to set many unrealistic targets.
For instance, it wants African states to establish sub-
regional clearing and payment arrangements not later than
the end of 1984. Again, it calls for the establishment
of an African Payments Union before the end of the 1980s.
Significantly, the realisation of many of the Plan's
major objectives is dependent on the fulfilment of these
unrealistic targets.

Another shortcoming of the Plan is the absence of
any machinery to coordinate and monitor its implementation
by member states. The establishment of such a body would
have gone a long way to compensate for the non-mandatory
nature of the Plan. This weakness seems to have been
spotted out by the Heads of States of the OAU during the
June 1983 summit in Addis Ababa. At that meeting African
leaders called for biennial progress reports by the OAU
Secretariat in close collaboration with the ECA. Even
then, this is a far cry from the establishment of a body
that would monitor the implementation of the Plan. Apart
from that, it is doubtful whether the OAU Secretariat,
as it is presently constituted, has the resources to carry
out such a task.

The successful implementation of the Plan requires
huge sums of money and sophisticated technology, both of
which are notoriously scarce in Africa at the moment.
For instance, the Plan envisages an expenditure of $22
billion between 1980 and 1985 for the realisation of its
Food Programme alone. Half of this money is expected to
come from the international community, i.e., the International Fund for Agricultural Development and the World
Food Programme. This raises a number of vital issues.
First, are Africa's leaders so sure that the international
community will readily pump such a huge sum of money to
Africa simply because they say Africa needs it?
Experience has not borne this out, and there is nothing
to indicate that the attitudes of the international
community towards the continent have significantly
changed. Second, even if the international community
would readily provide the needed funds, that itself
raises the whole question of the continuing dependence of
the continent on extra-African powers. It is rather
ironic in that regard that a blueprint which seeks to make

Africa self-reliant in food, energy and other vital areas
is so heavily dependent upon the generosity of foreign
powers for finance and technology for its realisation.
And finally, where will Africa get its own share of the
$22 billion that is needed to implement the programme?
This question becomes more pertinent when viewed against
the background of the alarmingly poor financial situation
of the continent. Almost half of the countries of Africa
are classified amongst the least developed countries of
the world. The current balance of payments deficits of
most African countries are staggering. Indeed by the late
1970s the external debts of Africa's least developed
countries had already soared to over $25 billion. The
situation becomes more serious when oil-producing countries like Nigeria, which were thought to be immune to
balance of payments problems in the 1970s, are now having
serious liquidity problems. In mid-1983, Nigeria's
external debt alone was $6.7 billion and it had to
approach a consortium of western banks and the IMF for a
rescue operation."[21]
    Besides the above weaknesses, the success of the
Plan is dependent on the cooperation of African leaders.
Unfortunately this cannot be taken for granted given the
variegated political map of the continent. Furthermore,
personality incompatibility and other idiosyncratic factors have continued to hinder sub-regional and continental
cooperation both in the economic and political spheres.
The Lagos Plan of Action calls for a lot of selfless
sacrifices on the part of African leaders if it is to
succeed. It is doubtful, however, whether these leaders
have the dedication and the political will to make such
sacrifices.

NOTES

    1. See OAU Doc. CIAS/Plen.2/Rev.2; and CIAS Plen.3.
    2. OAU Doc. CM/316, 1970.
    3. OAU Doc. AHG/67(Part 11), p. 7.
    4. OAU, African Declaration on Cooperation,
Development and Economic Independence, CM/ST.12(XXI), 1973.
    5. Ibid., pp. 12-14.
    6. For details, see Olusola Ojo, Afro-Arab
Relations (London: Rex Collings, 1982), Chapter 5.
    7. Ibid.
    8. Ibid.
    9. Ibid.
    10. For details, see Sola Ojo, "The Arab Israeli
Conflict and Afro-Arab Relations," in Timothy M. Shaw and
Sola Ojo (eds.), Africa and the International Political
System (Washington: UPA, 1982), pp. 152-156.
    11. For details, see Olusola Ojo, Afro-Arab
Relations.

78

12. See Colin Legum (ed.), <u>Africa Contemporary Record</u> (ACR), Vol. 9, 1976-1977 (London: Rex Collings, 1977), p. A74; and also <u>ARB</u> (Economic and Technical Series), 15 November-14 December 1976, p. 4087.

13. <u>What Kind of Africa by the Year 2000?</u> (OAU/ International Institute for Labour Studies, Geneva, 1979), p. 14.

14. Ibid., p. 17.

15. Ibid., p. 18.

16. OAU Doc. AHG/ST.3(XVI).

17. OAU, <u>Lagos Plan of Action for the Economic Development of Africa, 1980-2000</u> (Geneva: International Institute for Labour Studies, 1981), p. 7.

18. Andrew M. Kamark, "Sub-Saharan Africa; Economic Profile," in H. Kitchen (ed.), <u>Africa: From Mystery to Maze</u> (Lexington, Mass.: Lexington, 1976), p. 68.

19. <u>West Africa</u>, 26 May 1980, p. 919.

20. OAU, <u>Lagos Plan of Action</u>, pp. 83-84.

21. <u>Daily Sketch</u> (Ibadan), 27 June 1983.

5
The OAU and Human Rights

Concern for the protection and promotion of human rights predated the formation of the OAU in 1963. However, despite the existence of the UN Universal Declaration of Human Rights, the OAU Charter only made passing references to the issue of human rights in its preamble and in Article II. Moreover, African leaders gave a state-centric interpretation of human rights. To them the UN Declaration was interpreted as merely a means of promoting "peaceful and positive cooperation among States." Yet, while most of the clauses in Article III, which states the principles of the Organisation, are mere recapitulation of the UN Charter, the African leaders made no further references to human rights.

This neglect in 1963 is hardly surprising. The major preoccupation of the African leaders was how to enhance the security of the fragile state system which they had then just inherited from their erstwhile colonial masters. Already in this period some independent African states were seriously threatened by internal as well as external forces. Indeed, many of the African leaders that gathered in Addis Ababa in May 1963 were already under threat from internal opposition groups. The Congo Crisis and the assassination of Sylvanus Olympio of Togo in a *coup d'etat* shortly before the founding conference of the OAU graphically illustrated this dilemma. Furthermore, the experience of African leaders during the colonial era did not inculcate in them respect for human rights. Many of them had served prison terms for daring to criticise the colonial policies. Besides, they inherited the colonial instruments of repression: the police, the intelligence services, the army, and the prisons.

Arising from the above factors is the absence of any commitment to the promotion of economic, social, and cultural rights of the Africans. As we have already made clear in Chapter 4, the attention of the OAU in the first seventeen years of its existence was almost exclusively devoted to political issues. And even if the OAU had been committed to the promotion of the above rights, member states were ill equipped both financially and technologically to support such a commitment.

HUMAN RIGHTS: THE AFRICAN EXPERIENCE

The African record on human rights issues had been a bad one. The history of most post-independence African states is strewn with gross violations of human rights. Although at independence most of the states inherited a multi-party state system, this was, however, quickly dismantled to give way to one-party systems or military dictatorships. This was usually accompanied by violent suppression of opposition elements. The experience of the supposedly multi-party states has not been different either. Mass arrests, detention without trial for long periods of time, mock trials and secret execution of opposition elements have been common place in Sekou Toure's Guinea; Egypt, Sudan, and Zaire, according to Amnesty Report, to name a few, have been guilty of these crimes.

In most African countries, freedom of the press and speech as well as freedom of assembly are enshrined in their constitutions. In practice, however, these fundamental rights are denied the citizens. Newspapers are, in most cases, government owned, and only reflect government opinion on domestic and international issues. There are spirited attempts to suppress opposition newspapers. There are frequent arrests and detention without trial of journalists; similarly, kangaroo press laws aimed at stifling opposition press are passed. There have been violent and arbitrary closures of opposition press. This was the experience, for example, of the *Nigerian Tribune* towards the end of the Nigerian First Republic. Press harrassment and intimidation through charges of sedition brought against editors have also featured in the Second Republic. The experience in Sierra Leone is even more shocking. In 1970, gunmen believed to be government soldiers opened fire very early in the morning on the opposition press, *The Express*, with the main objective of killing the editor. Earlier in the same year, another press, *Freedom*, owned by the newly-formed United Democratic Party, a break away wing of the All Peoples Congress, was attacked and the machines destroyed by government agents.[1] As a result both papers were killed. And in July 1983, the controversial Newspaper Amendment Act, tagged "Killer Bill," was implemented. According to the Act, only four newspapers--the government-owned *Daily Mail*, the ruling party bi-weekly, *Weyone*, a sports weekly, *Progress*, and the supposedly independent weekly *Flash*--would be allowed to publish with effect from 1 July 1983.[2] Consequently eleven other newspapers were banned. In the Ivory Coast in 1980, eighteeen journalists and technicians working at Television Ivoirienne and

Agence Ivoirienne de Presse were arrested, forcibly conscripted into the army, and sent into the military camp at Seguela.[3]

As would be expected given the preponderance of single party states, elections are not only not free and fair, but are often accompanied by violence and loss of life and property. Sadly enough, even in the few states that still practice a multi-party system, the experience is the same.

The perennial problem of military *coups d'etat* in Africa has also compounded the problem of human rights violations. Military coups have been followed by gross violations of the political and civil rights of the erstwhile rulers and their supporters. The aftermath of the first Jerry Rawlings coup in Ghana in June 1979 saw the swift execution of three former Heads of State without proper judicial procedure. Again, the April 1980 coup in Liberia was accompanied by the executions of thirteen top government functionaries.

Besides successful *coups d'etat*, there have been numerous abortive attempts to topple both civilian and military regimes in Africa. Each of these attempts was followed by witch hunts, arrests, detention and executions of suspects and political rivals. In 1975, for example, all the major political rivals of Siaka Stevens of Sierra Leone were eliminated following allegations of an attempt to overthrow his government. In the Gambia between ten and twenty detainees arrested at the time of the abortive coup in July 1981 died of suffocation while in police custody.[4] Among the dead were Femi Djeng, a well-known journalist, and Koro Sallah, brother of the leader of the abortive putsch. Again in the Sudan, numerous coup attempts had been followed by detention and executions of suspects.

Finally, the tenuous nature of the social and political fabrics of some African states did not survive the departure of the colonial powers for too long. In Burundi, ethnic rivalries between the Hutus, who constitute about 85 percent of the country's population, and the Tutsis, who dominate the government, led to the massacre of an estimated 300,000 Hutus in April 1972.[5]

THE LONG ROAD TO AN AFRICAN HUMAN RIGHTS CHARTER

The OAU remained indifferent to the many instances of human rights violations on the continent. More than four million Africans are officially listed as refugees—more than half of the world's total—and the number of Africans killed by their own governments since the early 1960s is probably more than a million.[6]

Apart from the reasons which we have already advanced, the OAU's attitude was partly a result of the fact that the Organisation itself is by and large a league

of heads of states and governments whose major pre-
occupations are the same: self-preservation, regime
security, and the protection of the territorial integrity
of their states. Most of the human rights violations were
carried out in the name of the protection of at least one
of the above interests. Not surprisingly, there were no
loud protests from other leaders.

Furthermore, some provisions of the OAU charter are
themselves a major obstacle to the promotion of human
rights on the continent. Article III of the Charter
categorically forbids interference in the internal affairs
of states by members of the Organisation. This provision
has many times in the past been invoked to prevent other
states from raising human rights issues on the continent.
For instance, the civil wars in Ethiopia, Nigeria, and the
Sudan and the genocide in Burundi and Rwanda, in which
millions of innocent African lives have been lost, were
never discussed within the context of protecting or promoting human rights. In the Nigerian case, the OAU was
concerned with how to restore the unity of the country
(i.e., its territorial integrity), and Burundi was only
discussed within the context of the refugee problem.

However, the OAU had become more concerned with human
rights issues since 1979. Thus "new" awareness was brought
about partly by external pressures and partly by developments within some African countries, mainly Uganda,
Equatorial Guinea, and the Central African Republic. The
human rights crusade of the then American President,
Jimmy Carter, brought human rights to the forefront of
international politics. As a result, both the United
States and other Western countries became critically out-
spoken on human rights violations not only in the socialist
countries as had hitherto been the case, but also in the
Third World. Besides, an increasing number of these
countries began to link their foreign aid programmes with
human rights records in the prospective recipient
countries. In 1980, the EEC suspended its aid to Liberia
on the grounds of human rights violations following the
coup and the assassination of President Tolbert.[7] Earlier
in April 1978, the United States suspended aid to
Ethiopia for alleged human rights violations by the
Mengistu regime. About three months later, Canada and
Sweden announced that they were also suspending assistance
to Ethiopia for similar reasons.[8]

The emergence of three contemporary tyrants/dictators
on the continent--President-for-life Idi Amin of Uganda,
President-for-life Francisco Macias Nguema of Equatorial
Guinea, and self-crowned Emperor Jean Bokassa of the
former Central African Empire--compelled the OAU to turn
its attention to the very serious human rights violations
in Africa. In Uganda, the last years of Idi Amin's seven-
year reign of terror saw the deaths of thousands of his
political opponents. Amnesty International puts the
figure at 300,000.[9] In Equatorial Guinea, Nguema's

atrocities throughout his ten-year reign of terror led to the flight of almost half of the 3,054,000 inhabitants of the country. Emperor Bokassa's reign in the Central African Empire was not better either. His notorious personal participation in April 1979 in the murder of about 100 school children for defying his order to wear school uniforms was just one in a long list of atrocities he committed before he was overthrown in 1980.[10]

This particular incident played a major role in making human rights violations by African states a major issue at the 16th summit in Monrovia in July 1979. The killings led to the setting up of a panel of judges from Ivory Coast, Liberia, Rwanda, Senegal, and Togo by the Franco-African Summit meeting in Rwanda in May 1979 to investigate the alleged involvement of the Emperor. Their 137-page report, which they presented to their respective governments on the eve of the Monrovia summit, concluded: "His responsibility was involved. His presence at the sites is proven, his participation is quasi-certain."[11] Since one of the judges came from Liberia, which hosted the 1979 summit, the report was bound to influence the position of that country. Not surprisingly, for the first time in its sixteen years of existence, President Tolbert in his opening address admonished the OAU for its unacceptable human rights record.[12] He criticised the unwillingness of his fellow African leaders to speak out against human rights violations on the continent. The Liberian leader observed that the OAU's inviolate principle of non-interference "has become an excuse" for African leaders to remain silent on clear instances of human rights violations. He warned his colleagues that Africa's "sovereignty is empty" if it produces "poverty and brutality" and recommended a revision of the OAU Charter to protect fundamental human rights.

This speech set the pace for the discussion of other issues, including the Tanzanian-backed ouster of Idi Amin. Nyerere justified his action at the summit on the grounds that Amin's regime had been a catalogue of gross violations of the human rights of the Ugandan people. In fact, Idi Amin's Uganda, along with Cambodia and the Central American republics, were the worst countries in terms of human rights protection on Amnesty books in the 1970s.[13] Apart from discordant notes by Nigeria and Sudan, the other African leaders supported Nyerere's position. Furthermore, there was a consensus among the African leaders that it was high time the Organisation took a firm stand on the issue of human rights in Africa. Thus a proposal by the President of Senegal to have a charter of human rights was unanimously adopted by the summit. The summit then requested the Secretary-General to convene a restricted meeting of experts to prepare a draft charter on Human and Peoples' Rights. At the end of the summit, Peter Onu, who acted as the OAU spokesman, put it this way: "We felt it was high time we had such a charter. We cannot be talking about the denial of human rights in certain parts of Africa (where whites retain power) if we don't accept the same standards ourselves."[14]

The committee of experts met in Dakar from November 28 to December 8, 1979 to draw up a draft charter. The Secretary-General enjoined the experts that in drawing up the charter they should take cognisance of the "African concept of Human Rights." The objective was to make the proposed charter distinct from other conventions already adopted in other regions. He said it was therefore essential for them to:

(a)   give importance to the principle of non-discrimination;

(b)   lay emphasis on the Principles and objectives of the OAU as defined in Article 2 of the Charter of the Organisation of African Unity with particular reference to respect for the sovereignty and territorial integrity of each State and for its inalienable right to independent existence; and to absolute dedication to the total emancipation of African territories which are still dependent;

(c)   include Peoples' Rights besides individual rights;

(d)   determine the duties of each person towards the community in which he lives and more particularly towards the family and the state;

(e)   show that African values as well as morals still have an important place in our societies; (and)

(f)   give economic, social and cultural rights the place they deserve.[15]

The draft prepared by the experts was considered six months later at a conference of OAU Ministers of Justice in Banjul, Gambia. However, the Ministers failed to adopt the draft charter. They approved only eleven of the sixty-three draft articles during the seven-day conference. Throughout their discussions, emphasis seemed to have been concentrated on ensuring that the final product is distinctively "African." This was reflected in some of the amendments made to the original draft. For example, many delegates argued that in Africa Man is inseparable from the group. Consequently individual rights could only be explained and justified by the rights of the community. They therefore advocated for the inclusion of peoples' rights as distinct from human rights.[16] Furthermore the Ministers added an entirely new paragraph to the draft preamble of the experts which read: "Taking into consideration the virtues of their historical traditions and the values of African civilisation which should inspire and characterise their reflections on the concept of human and peoples' rights."[17]

The second and final meeting of the Ministers of Justice was convened in January 1981 in Banjul to consider the remaining articles of the draft charter. During the conference four new articles were added to the original draft. One of these is Article 57 of the Charter which stipulates an additional condition to be fulfilled before any issue referred to the proposed Commission on Human and Peoples' Rights is considered by the commission. The new article demands that the Commission must refer all communications brought to it by an aggrieved party to the attention of the state concerned before any substantive consideration of the issue by the Commission.

After fourteen days of deliberations, the Ministers finally adopted the draft on January 19, 1981. However, Burundi, Ghana, Kenya, Tanzania, and Zambia made reservations on some of the Articles. Most of them expressed reservations on paragraph 3 of Article 45. Besides, Tanzania also expressed general reservations on all articles of the Charter relating the Commission with the Assembly of Heads of State and Government.

There are many provisions in the draft charter which seem to make the Commission subservient to the Assembly of Heads of State and Government. It is also significant to note that an attempt to include in the draft a provision for the establishment of an African court to judge crimes against mankind and violations of human rights was rejected on the ground that "it was untimely."[18]

The amended draft charter was adopted by the 18th summit of the OAU in Nairobi in July 1981. However, as stipulated in Article 63 the Charter will not come into force until three months after the Secretary General of the Organisation has received the instruments of ratification or adherence of a simple majority of the member states of the OAU. As of October 1983 only ten states have ratified the charter.[19]

THE AFRICAN CHARTER ON HUMAN AND PEOPLES' RIGHTS

Apart from the pious intentions of African leaders in protecting and promoting human rights contained in its Preamble, the Charter is divided into three broad sections. Part I spells out the rights and duties of the individual and peoples. Part II deals with the establishment of the African Commission on Human and Peoples' Rights, its mandate and procedure. Finally, Part III deals with general provisions concerning the ratification, adherence to, and amendment of the Charter.

The Charter sets up an African Commission on Human and Peoples' Rights composed of eleven individuals elected by secret ballot by the Assembly of Heads of State and Government from a list of persons nominated by member states. The main functions of the Commission are to

promote and ensure the protection of Human and Peoples'
Rights in the continent. This it can do in a variety of
ways. First, it can undertake studies and public
enlightenment exercises. Second, it can lay down
principles to guide member states in legislating on human
and peoples' rights. Third, it can interact with African
and international institutions. Finally, the Commission
can hear complaints from member states and individuals
or organisations, and tries to solve such disputes. However, if conciliation fails, the Commission then makes a
report of the case to the AHG.

Almost two and a half years after its unanimous
adoption by African Heads of State and Government only
fifteen of the fifty-one member states have so far ratified the charter. This is just one of the obstacles in
the way of any institutionalised protection and promotion
of human rights in Africa. Even if and when the Charter
is ratified by the required number of states, a whole lot
of problems will still stand in the way of the realisation of the noble objectives of the Monrovia summit which
called for an African Charter on Human Rights.

The Charter is evidently too state-centric. There is
so much emphasis on the role of the state that it is
questionable whether individual rights will be adequately
protected. In fact, the whole issue of individual rights
was expressly made subservient to peoples' (or states')
rights as we have earlier indicated. Under Articles 47
to 54 of the Charter, communication concerning human
rights violations in any part of Africa can in the first
place only be made by another state, not by an aggrieved
group or individual. Even then, such communication is
addressed to the state accused of violations, and not
directly to the Commission. In fact, the Commission does
not get involved until three months after such complaint
and that is if the issue has not been "settled to the
satisfaction of the two states involved through bilateral
negotiation or by any other peaceful procedure." Even
when communications concerning violations are sent
directly to the Commission as provided for in Article 49,
the Commission cannot act on such complaints until it is
satisfied that all local remedies have been exhausted
(Art. 50). Besides, when the Commission acts, the
objective is "to reach an amicable solution" between the
two states. Under such circumstances it appears the
Commission will be settling inter-state disputes rather
than being a watchdog on human rights violation. This is
a poor contrast to the practice of the European
Commission of Human Rights.[20]

The provisions made for petition by non-state actors
especially individuals, are to say the least grossly
inadequate. First, such petition will be entertained
only if it secures a simple majority of the eleven-man
Commission (Art. 55.2). This contrasts unfavourably with
the situation in Western Europe where the right of

individual petition has been accepted by most state
parties to the European Commission of Human Rights. Given
the fact that members of the Commission are in the final
analysis nominees of their national governments, this
requirement could easily open them to undue political
pressure or influence, thus negating the spirit of the
Monrovia summit.

Second, even when the Commission is predisposed to
entertaining a petition from an aggrieved individual,
such petition could be adversely jeopardised by the provisions of Article 56. Among other things, the Article
stipulates that petitions should be compatible with the
Charter of the OAU. This is vague and can easily be used
to forestall consideration of petitions by the Commission,
no matter how serious the allegations are. It is not
stated which aspect of the OAU Charter a petition has to
comply with. It is strange that such a clause could even
be included in the Charter given the fact that the OAU is
essentially an inter-state organisation. Besides,
paragraphs 3 and 5 of Article 56 impose two more hurdles
which a petitioner has to scale before he could be heard
by the Commission:

(i)   he must not write in disparaging or insulting
      language directed against the State concerned
      and its institutions or to the Organisation
      of African Unity,
(ii) he must have exhausted local remedies.

It is undoubtedly alarming to think that the Commission
would have rejected a petition against gross violations
of human rights by regimes like those of Idi Amin,
Bokassa, or Nguema, simply because the petition was
written in "disparaging or insulting language." Besides,
given the experience in most African states, the charter
is expecting too much by saying that a petitioner must
have exhausted "local remedies" before recourse to the
Commission. Even where there are such provisions on state
statutes, some leadership has so much grip on the state
apparatus--the police, army and the judiciary--that it
will be foolhardy to expect any individual to attempt
to use these institutions to get redress against an in-
justice committed by the leadership. Such an attempt
may lead the aggrieved petitioner to more serious
difficulties, such as perpetual incarceration or even the
loss of his life.

Third, even when the Commission is satisfied that one
or more communications to it "reveal the existence of a
series of serious or massive violations of human and
peoples' rights" the Commission can not "undertake an in-
depth study of these cases" without the prior approval
of the Assembly of Heads of State and Government of the
OAU: The only thing the Commission can do is to draw
the attention of the OAU's Assembly to these cases. The

Assembly may then request the Commission to investigate the cases. However, given the fact that the Assembly is composed of Heads of State and Government, and given the fact that the decisions of the Assembly are often influenced by personal friendships and shared ideology of these leaders, it is possible that the matter may die in the Assembly.

Closely related to the above problems is the lack of sanctions by the Commission. Unlike its European counterpart, which can refer cases straight to the European Court of Human Rights and can ultimately expel a member from the Organisation, there is no provision for an African Court of Human Rights or expulsion. Indeed, as we indicated earlier, a suggestion to have a court was flatly rejected by the OAU Ministers of Justice. The Commission can only submit a report of its findings to the Assembly of African Heads of State and Government. At best, the only probable deterrent is perhaps the possibility of investigation, publicity, and condemnation of the offending state by the AHG. This is, however, not automatic. As Article 59(i) states, all measures taken within the provisions of the Charter "shall remain confidential until such a time as the Assembly of Heads of State and Government shall otherwise decide." As experience with Idi Amin, Bokassa, and Nguema has shown, it is very doubtful whether the threat of, or actual publicity could have deterred them from further violations.

Another worrying aspect of the Charter is its concept of "peoples' rights." Too much emphasis on "peoples' rights" might negate the protection of individual rights. Undoubtedly it could be used to suppress the rights of individuals who might reject generally accepted morality. Besides, it opens the Charter to different interpretations by states and individuals according to their ideological convictions.

One significant feature of this Charter which may lead to some difficulties in its implementation and which could indeed forestall the promotion of individual rights is the emphasis placed on "duties." Some of the provisions under Duties are not only vague but are also unrealistic and retrogressive. At best they are mere pious hopes which would have no practical effect. Article 29(1), for instance, imposes on the individual the duty to "preserve the harmonious development of the family and to work for the cohesion and respect of the family, to respect his parents at all times, to maintain them in case of need." At worst, the provisions are oppressive and could in fact be used by the state to curtail human and peoples' rights.

Like the Duties, some of the Rights are couched in such vague language as to make them meaningless. Notable examples are Article 16.1 and 24. Besides, the provision of Article 20.1 is not only vague but it is also in direct conflict with some provisions of Article 29. While the

OAU Charter patently recognises the inalienable rights to
self-determination of peoples under colonial or racist
minority regimes, it does not accord such a right to
communities within sovereign independent African states.
However, the Article makes a blanket statement to the
effect that all peoples "shall have the unquestionable
and inalienable right to self-determination. They shall
freely determine their political status." Do the
Eritreans in Ethiopia, the Ewes in Togo, the Somali in
Kenya and Ethiopia, for example, have an inalienable
right to self-determination? Perhaps such confusion
and misinterpretation could have been avoided if there
had been a clear definition of what "peoples" mean in the
context of the Charter. But the Committee of Experts
that drafted it deliberately avoided the issue of
"definition of such notions as 'Peoples' so as not to end
up in difficult discussions."

It is pertinent to emphasize that even the enjoyment
of some of the rights enshrined in the Charter is not
automatic: it is made dependent on the fulfillment of
certain conditions laid out in the Charter. For example,
the Charter provides for the right of the individual "to
express and disseminate his opinions within the law"
(Art. 9.2) and the "right to free association provided
that he abides by the law" (Art. 10.1). Again, the
exercise of the individual right to "assemble freely with
others" is subject to "necessary restrictions provided for
by law in particular those enacted in the interest of
national security, the safety, health, ethics and rights
and freedoms of others" (Art. II). These conditions fail
to take into cognisance two vital considerations. First,
the proliferation of one-party or non-party political
systems. And second, is the hypersensitivity of all
African leaders to opinions and ideas other than those
which they themselves believe in. It is therefore
possible for the individual and even peoples to be denied
these rights on the whims and caprices of their national
leaders under the cloak of violation of the law and in
the interest of "national security." It is therefore
ominous that in Liberia, Commander Samuel Doe dismissed
his Minister of Foreign Affairs, Boima Fahnbulleh, for
expressing opinions which are not consonant with his
own.[21]

Even if one were to ignore these formidable hurdles
to the protection and promotion of human and peoples'
rights in Africa, it is still possible for states to
invoke the non-interference clause of the OAU Charter to
prevent the Commission from performing its duties. Given
the lack of sanctions by the Commission and the OAU itself, there is nothing to discourage any determined
African leader from preventing the Commission from
investigating alleged violations of Human and Peoples'
Rights in his country.

Finally, the Charter seems to have highly politicised the whole issue of human rights in Africa. As we have already pointed out, too much power has been invested in the AHG. Besides, the Commission is an organ of the OAU. The Secretary-General of the OAU appoints the staff and services necessary for the effective discharge of the duties of the Commission and the cost of its staff and services is to be borne by the OAU. It is possible for a state accused of violating the Charter and its friends to suspend all contributions to, and even participation in the activities of the OAU. From the experience of the Organisation in 1982 over the holding of its 19th Summit in Tripoli, such a situation could lead to the complete paralysis of the OAU. More intriguing is the inclusion of "Zionism" in the African Charter. The only plausible explanation for its inclusion is perhaps to soothe Arab sensibilities.

Nonetheless, the Charter is a significant step towards institutionalised protection and promotion of human rights in the continent. It has highlighted issues such as the rights of women and children and the economic and social duties of the state to its citizens, which have hitherto not been given due attention. However a lot of political goodwill and the determination to promote and protect Human and Peoples' Rights are needed from African leaders if this initial effort is not to miscarry, given the serious loopholes already highlighted in the African Charter.

NOTES

1. One of the co-authors of this book personally witnessed the incidents.
2. West Africa, 18 July 1983, p. 1687.
3. See Amnesty Annual Report for 1982. A brief summary of the Report is contained in West Africa, 15 November 1982, pp. 2953-5.
4. Ibid., p. 2954.
5. Colin Legum (ed.), ACR, Vol. 5, 1972/73, p. B125.
6. This is in fact a conservative estimate given the enormities of regimes like Amin's, Bokassa's and Nguema's alone.
7. Amadu Sesay, "Le coup d'etat du Liberia: Facteurs internes et effets regionaux" in Politique Africaine, Vol. 11, No. 7, September 1982, pp. 91-106.
8. Olusola Ojo, "Ethiopia's Foreign Policy Since the 1974 Revolution" in Horn of Africa, Vol. 111, No. 4, 1980/81, p. 8.
9. Los Angeles Times, 21 July 1979.
10. For details of this particular sordid incident, see Jonathan Power, Amnesty International: The Human Rights Story (Oxford: Pergamon Press, 1981), pp. 82-89.

11. Agence Presse (Paris), 6 August 1979.
12. Address by H. E. William R. Tolbert, Jr.--to the 16th Summit meeting of the OAU on 11 July 1979 (Monrovia: Government Printer, 1979).
13. Power, Amnesty International, pp. 44-70, 82-89
14. Los Angeles Times, 21 July 1979.
15. Report of the Secretary-General on the Draft African Charter on Human and Peoples Rights, CM/1149 (XXXVIII), p. 1.
16. Ibid., p. 3.
17. Ibid., p. 5.
18. Ibid., p. 26.
19. The ten countries are Congo, Gambia, Guinea, Liberia, Mali, Nigeria, Rwanda, Senegal, Togo, and Tunisia.
20. For details see Michael Akehurst, A Modern Introduction to International Law (London: George Allen and Unwin, 1982), 4th Edition, pp. 78-79.
21. West Africa, 11 July 1983, p. 1592.

6
Conclusion: The Future of the OAU—Analysis and Practice

The OAU has over the last two decades come to be accepted by African and non-African leaders as the most important continental political organisation. It has become widely accepted as the forum to which African issues should first be referred before recourse to any other international organisation. This partly explains why the Organisation has been able to survive all the crises that have threatened its existence throughout this period. This is remarkable, considering the poverty of the Organisation itself in terms of financial, material, and human resources and the diversity of the political, ideological, economic, and cultural structures of its fifty-odd member states.

In this conclusion we examine this heterogeneity as well as Africa's homogeneity and analyse the continent's constraints as well as its resources. Following the organisation of the volume, we overview the related issues of decolonisation, conflict control, development strategies, and human rights and then proceed to preview these until the end of the present century, in addition to identifying possible optimistic and pessimistic scenarios. We conclude by considering the intellectual as well as political implications of Africa's futures: the next fifteen to twenty years.

One area where the OAU has made significant impact in its first twenty years is decolonisation. At its inception in 1963, there were only thirty-three independent states in Africa. However, by 1983 all African territories had become independent except Namibia and South Africa, where the problem of apartheid is still far from being solved. Through its moral, political, diplomatic, financial, and material support to the liberation movements, especially those in Rhodesia and the former Portuguese territories, the OAU not only made decolonisation an active issue in world politics but also gave the liberation struggles continental and global legitimacy.

Nonetheless, as we have shown in Chapter 2, the OAU's efforts in the sphere of *decolonisation* have not been the decisive factors responsible for change in the colonial territories. The OAU does not have a liberation army. Moreover, its material and financial support to the liberation movements was very minimal when compared with similar assistance from socialist states. It is significant to note that even the record of the Organisation in its attempt to unify rival liberation movements is a poor one, except perhaps in Zimbabwe, where the OAU through the efforts of the Front-Line States (FLS) was able to bring together, at least for a while, the two most important movements--ZAPU and ZANU--under the umbrella of the Patriotic Front.

Another area where the OAU has been very active is in *inter-African conflict control*. During the period under review the OAU attempted to settle every major conflict that broke out among its members. On several occasions its intervention prevented the internationalisation of African disputes. Since its founding in 1963 no inter-African dispute has been taken to the UN, except for the Chadian-Libyan dispute in August 1983 over alleged Libyan interference in internal Chadian affairs. Even then, the Security Council referred the issue back to the OAU. Besides, the existence of the OAU has provided disputants with a forum from which they could argue their case and thereby "let off steam." That way the Organisation has been able to put some disputes "on ice" while finding a lasting solution to them.

Nevertheless, the OAU has not been able to find definitive solutions to any of the major inter-African conflicts. The Ogaden problem has remained a running sore for the OAU since its inception in 1963. So have the disputes in Chad and Western Sahara, which have plagued the Organisation since the early 1970s. Several factors are responsible for this limited achievement. Among these are, first, the absence of a standing peace-keeping force. Unlike the UN, the OAU, as demonstrated in its peace-keeping fiasco in Chad in late 1981 to 1982, can neither raise nor finance a peace-keeping force to separate combatants in the battlefield. And second, there is the absence in its Charter of mandatory sanctions that would compel disputants to reach a peaceful settlement.

Although one of the rationales for creating the OAU was to harmonise *development strategies* of member states so as to promote the economic development of the continent, the Organisation neglected development issues for most of its first two decades of existence. Its attention was for most of this period devoted to political issues such as decolonisation and inter-African conflicts; most of the African states were much more concerned with political rather than economic survival. In retrospect, this is surprising, given the poor economic performance of these

states and the continual gloomy predictions of the ECA
and the World Bank. Ironically, political stability can
hardly be sustained in an atmosphere of stagflation,
sluggish and/or zero growth, mounting foreign debts and
the worsening food situation.
     It took the oil crisis of 1973-1974 to shake the OAU
into giving more collective attention to the issue of
economic development. The result has been a flurry of
conferences, strategies, and plans all aimed at reversing
the gloomy economic trends on the continent. So far,
however, all these have not made any significant impact
on the economic performance of member states either
individually or collectively.
     One other area that had hitherto been neglected but
that has now been given some attention, albeit belatedly,
is *human rights*. Although the history of post-independence
Africa is dotted with atrocities committed against
individuals and/or communities by governments, the OAU
did nothing in the direction of promoting and safeguarding human rights on the continent. The excesses of Idi
Amin, Jean Bokassa, and Macias Nguema at a time when the
protection and promotion of human rights had become a
major issue in world politics spurred the OAU into action.
By 1981 the OAU had come up with a Charter of Human and
Peoples' Rights. Unfortunately, however, the Charter has
not yet come into force. Moreover, it is defective in
many respects, as we have highlighted in Chapter 5.

PROSPECTS FOR THE FUTURE: POLICIES

     It is generally agreed that the OAU has over the last
two decades been preoccupied with the issues of decolonisation and political stability in Africa. These two
issues have also been the most cohesive factors that have
held the Organisation together. All African territories
except Namibia and apartheid South Africa have come under
black majority rule since 1963. The Namibian and South
African cases have exposed the inadequacy of the OAU's
strategy on decolonisation. For the OAU to continue to
be relevant in the ongoing liberation struggles in
Southern Africa it would have to change its strategy.
It has been proved that too much reliance on the moral
conscience of the international community to bring about
the desirable changes in Southern Africa is a total
failure. OAU members will have to accept the fact that
any meaningful change in Southern Africa can only take
place through their own actions. This, however, calls
for a number of measures and sacrifices on the part of
African countries, leaders, and peoples. They have to be
totally committed to the liberation of Namibia and to the
elimination of apartheid in South Africa. This means an
increase in financial, political, and material assistance
to SWAPO, the ANC, and the PAC. African countries should
also put more political and economic pressure on the

Western countries and multinational companies that have
been sustaining the South African presence in Namibia
and apartheid in the Republic. They should mount a
renewed and effective diplomatic offensive aimed at
committing the international community to the eradication
of colonialism and apartheid in Southern Africa.

The relevance of the OAU in the decades ahead will
also depend upon how much the Organisation is able to
adapt itself to the needs of its members. The Organisation will have to pay more serious attention to the vital
issue of economic development. It is disappointing that
since the 1980 economic summit in Lagos there has not been
any effective follow-up. Yet the economic situation in
Africa continues to deteriorate. The OAU will need more
economic summits in future. More important, the Organisation should set up a committee to review and implement
the *Lagos Plan of Action*.

To be able to cope with Africa's political and
economic needs for the rest of the century, the OAU
machinery itself will have to be overhauled. Some of its
Charter provisions will have to be revised in line with
Africa's contemporary needs. There is a need in that
regard specifically to insert in a revised Charter the
promotion of human rights and economic development as one
of its purposes. Furthermore, the provision on non-
interference in the internal affairs of member states will
have to be re-examined with a view to making it possible
for member states to bring to the attention of the
Organisation and the Commission on Human and Peoples
Rights, developments within a state that run counter to
the principles and purposes of the Organisation. It is
also pertinent to mention here the need for sanctions
against member states that violate the Charter and
decisions of the Organisation. The blatant disregard
for the Organisation's Charter and decisions in the past
has led to ridicule of the OAU both inside and outside
Africa. This trend will be reversed if the OAU is seen
not only to bark but also to bite. Sanctions would thus
give teeth to the Organisation.

There is also a need for a reorganisation of the
Secretariat in order to strengthen and increase its role
in both the economic and political spheres. It should
have knowledgeable and competent economic staffs so as to
be able to play a more meaningful and direct role in
economic matters. The powers of the Secretary-General
should be enhanced to allow him a more direct role in
initiating political and economic policies. For instance,
he should have statutory powers to bring to the attention
of the Assembly of Heads of State or any other competent
body developments within member states that he thinks may
threaten the peace and security of the continent. Furthermore, he should be able to bring to the notice of relevant
bodies of the Organisation any action of its members that
he thinks may bring the Organisation into disrepute or

that runs counter to the purposes and principles of the
OAU Charter.

The performance of the OAU will be enhanced if the
proposal for a "mini-Security Council" is adopted. The
Council, which would be a standing body, could meet at
short notice to discuss developments in the continent.
It would, for example, be a forum that could be seized
with urgent political and economic problems that require
the immediate attention of the Assembly of Heads of
States. This would, no doubt, be an improvement on the
current practice of leaving everything until the annual
summits.

For the OAU to realize the ideals set out in the
Charter of Peoples and Human Rights a lot of improvement
will have to be made in the Charter. First, the Commission
on Human Rights will have to be made more autonomous vis-
à-vis the Assembly of Heads of States; this would enable
the Commission to discharge its functions without undue
delay and political interference. Second, the Charter
should provide for sanctions; otherwise violators may
treat the Charter and the Commission with contempt. As
with its European counterpart, there should be provision
for the ultimate sanction of expulsion of serious offenders
from the OAU. And finally, to facilitate the work of the
Commission, the Charter should be revised to provide for a
court in which violators of human rights would be tried.
Such a court should have the power to try not only
malevolent incumbent leaders but also those who might have
lost power but are still living either in their respective
states or in exile. To facilitate this, African states
should enter into extradition treaties with each other on
the one hand and with their former colonial metropoles
on the other, to make it possible for offenders to be
tried. Such an arrangement will make it difficult, if not
impossible, for future Idi Amins and Bokassas to get
sanctuary.

PROSPECTS FOR THE FUTURE: SCENARIOS

Making predictions about the future of the OAU is
fraught with many difficulties. First, the African
environment in which the Organisation operates is
characterised by political instability. Second, the
continent is currently experiencing the most serious
economic crisis since the dawn of independence in the
early 1960s, a crisis that has called into question the
very viability of the African state system. Third, is
the sheer number of countries that make up the OAU
(currently 50, or 51 if we add the SADR); these states
have different ideological orientations as well as modes
of development. And fourth, the African system is
riddled with personality conflicts that have in the past
adversely affected the work of the OAU and have in recent
times called the very existence of the Organisation into

very serious question.

Nonetheless, we still feel that an attempt can be made at projecting the OAU's futures. After all, such prognoses will not be done in an analytical vacuum. As proposed in this book, they will be based on both the objective and subjective circumstances of the Pan-African organisation. In short, then, we can conjecture scenarios for the OAU on the basis of extrapolations from its past and present experiences and performances.

First, the future of the OAU will to a large extent depend on the willingness of its members to meet their financial commitments to the Organisation. In the past, the member states have not been forthcoming with their membership dues and contributions to the Liberation Fund. For instance, in February 1984, Mr. Peter Onu, the OAU's interim Secretary-General, told the Council of Ministers in Addis Ababa that only $3.6 million had been paid out of the Organisation's 1984 budget of $23.6 million. Besides, Mr. Onu also disclosed that members were not meeting their obligations to the Liberation Fund which was $16.6 million in arrears.

The implication of this situation for the future of the OAU is clear: Unless OAU members are prepared to back up their rhetorical commitment to the Organisation with real resources, its objectives will not be realised. Besides that, refusal by members to fund the OAU could also have serious implications for its very existence. It is conceivable that the OAU's activities would one day grind to a halt as it would not be able to continue to operate on a shoestring budget.

We, however, do not see an immediate improvement in the Organisation's financial predicament. In fact, all the available evidence points to a bleak and precarious financial future for the Organisation. At the moment, the majority of its members are going through what we have already described as the most serious economic crisis since the early 1960s. Besides, there seems to be no immediate end to the world economic recession. This point is important because most OAU members have been incorporated into the world capitalist economic system. Unless there is economic recovery in the advanced industrialised as well as African countries, then, the fortunes of the OAU would remain dull. In short, we do not expect financial solvency for the OAU in the near future, perhaps not even by the year 2000.

This pessimistic scenario is of direct relevance to that of the future of decolonisation in the two remaining colonial and white supremacist enclaves in Africa-- Namibia and South Africa. Obviously, the ability of the OAU to prosecute the liberation wars in both territories successfully would to a large extent depend on the resources available to the Organisation. In that regard, we envisage reduced participation by the OAU in the liberation struggles now being waged in Southern Africa.

The implications of this scenario are (1) that the citizens of those countries/territories would be called upon to liberate themselves; (2) that the liberation movements in the two territories would have to look increasingly outward for necessary support for their wars--support would be financial, material, and even moral; and (3) that the OAU's role in the liberation of the remaining colonial strongholds would be peripheral, although this last observation is subject to a couple of important qualifications.

The first qualification is that OAU membership would consist essentially of three types of states: the sub-imperial states, the transformationist states, and the largely poor and agrarian states. If we conjecture an OAU dominated by the first group of states in the future, then the above scenario would tend to hold, that is, reduced OAU involvement in the last-ditch struggle to dislodge the colonialist and racist regimes in Southern Africa. This is because the sub-imperial states would tend to opt for a decolonisation strategy in Southern Africa that would be in consonance with their status and roles at the core of the African periphery. They would support the transfer of power in South Africa proper, for instance, into the hands of a moderate capitalist African elite. Thus, they would advocate a negotiated settlement to the conflict. Such a scenario plays down armed con-frontation as a means of achieving the OAU's/Africa's objective in Namibia and South Africa. Indeed, the sub-imperial states would view a South Africa under moderate African leadership not only a natural ally but also as a source of scarce capital and technology for their own economies. In this type of situation, the OAU could even team up with the capitalist West to pressure the liberation movements in South Africa and Namibia into a negotiated settlement. It is conceivable also that resistance from the liberation movements in those territories could easily lead to the pruning down--and perhaps the eventual termination--of OAU support to them.

The alternative decolonisation scenario is based on the premise that the OAU would be dominated by trans-formationist/radical African states with an appreciable amount of resources at their disposal. These states, according to our scenario, would try to decouple from the Western-dominated world economic system so as to reduce their and the OAU's dependence on external powers. We expect such a group of states to favour very strongly the emergence of radical black majority regimes in South Africa and the Namibia with a *materialist* bias. Under their leadership, we expect the OAU to play a more direct role in the liberation of the remaining colonial and racist enclaves in Southern Africa. Besides, the Organisation would also vigorously pursue the liberation wars, for which purpose it would openly seek support from the Eastern countries. In short, then, in the second

case we envisage a situation in which the OAU's members
would meet their commitments to the Liberation Fund and/or
a group of states would take it upon themselves to
finance the Fund to prosecute the liberation wars.

The futures of the OAU and conflict control in Africa
would very much again depend on which of the two major
types of states identified earlier actually materialises
and dominates. For instance, under an OAU that is over-
whelmingly dominated by transformationist states we would
expect a number of measures that would be aimed at
strengthening the Organisation's conflict control
machinery: (1) the creation of an African High Command
would be vigorously pursued; (2) more powers would be
given to the Secretary-General; and (3) stiffer penalties
would be laid down for states that flagrantly violate OAU
decisions relating to conflict control efforts.

The creation of an African High Command would enable
the OAU to mount swift and effective peace-keeping
operations in the continent. Such a capability would
enable the Organisation to nip in the bud not only
potential crisis situations but also to contain those
which have already flared into open confrontation among
its members. Again, limited executive powers for the
Secretary-General would greatly augment the activities of
the High Command. Thus, he would be able to take quick
measures to be aimed at settling intra- and inter-African
conflicts as well as acts of external military intervention in between sessions of both the Council of
Ministers and the AHG. The Secretary's initiatives in
that regard would be greatly boosted by the knowledge that
a party to a dispute that is reluctant to bring the dispute to a peaceful conclusion could be subject to
sanctions, which could in the first instance mean
suspension from the OAU, to be followed by dismissal
if the recalcitrance continued. It is conceivable that
if such sanctions had been available to the OAU, then
the Western Saharan disputes would have been settled--
or at least, Morocco would not have flagrantly defied
the OAU's peace initiatives.

Finally, the above scenario is predicated upon the
belief that a group of financially solvent radical states
would be quite prepared to make the financial sacrifices
needed to make the peace-keeping operations a success.
South Africa under a transformationist regime would play
a crucial role in the fulfillment of this scenario. After
all, the major objective of the radical states is to see
that the OAU is inward-looking, more self-reliant, and
capable of curbing undue external influences in the
continent.

On the other hand, it is also possible that the OAU
would be dominated by conservative states that would
closely collaborate with the capitalist world. Under
such circumstances, there would be an uneasy peace in the
continent enforced by what we can call a "Pax sub-

imperiania." The freedom of the smaller members of the
OAU would be sacrificed for peace that would secure their
markets for the sub-imperial states. The sub-imperial
states would accordingly divide the continent into spheres
of influence. Thus, the OAU would be unable to stem
external intervention in Africa, as the sub-imperial
states would collude with capitalist states to maintain
order in their client states or help their disgraced
colleagues back to power in their territories. However,
whatever order prevails in the continent at this time would
be at the expense of the freedom of the small OAU members,
who would become mere prey to the sub-imperial powers.

Unhappily, we do not believe that the OAU's present
development strategy would lead to a virile and self-
reliant Africa by the year 2000. On the contrary, we see
a continent that is increasingly dependent politically,
economically, and even militarily on external powers
because of its weak economic base. It is possible nonetheless that there would emerge relatively powerful
and radical transformationist states on the continent.
Such states would be opposed to the incorporation of Africa
into the fringes of the capitalist world system. They
would also strongly oppose what they would perceive as
the collaborative role of the sub-imperial powers with
respect to the capitalist states. In addition, transformationist regimes would try to redirect the OAU's
development strategy and would put more stress on self-
reliant economic development policies within a continental
framework. We thus envisage a radical overhaul of the
Lagos Plan of Action to remove all the neo-colonial
trappings it presently contains. Logically, we would also
expect the radical states to reject in its totality the
World Bank's *Agenda for Accelerated Development in Sub-
Saharan Africa*. If accepted in its present format, this
Agenda would hasten further Africa's incorporation into
the capitalist world economic system, a development that
would be contrary to the policies of the radical states.
With such a puritanical ideological approach or trend in
Africa's problems, we foresee a situation whereby the
conservative states would be forced to look for an
organisation of their own. Such an organisation would
later be used--with the help of external powers--to
challenge the dominant "pretensions" of the radical
states. Such an alternative is in fact what the
Francophones are trying to evolve, albeit very slowly,
at the moment.

Finally, if the OAU does not split along ideological
lines it could yet do so along racial ones. Because of
the continued disenchantment over Arab attitudes towards
aid, decolonisation, intervention in African states'
internal affairs, and the peripheral importance of the
Organisation to the Arabs when Arab interests are not at
stake, the black Africans may simply walk out of the OAU
to form their own organisation. This possibility should

not be considered a remote one in view of the fact that
such an idea has been floated at the OAU Secretariat
since 1977 and has recently been backed by at least one
of the Francophone African states.

As for the futures of the OAU and human rights in
Africa, we do not see any bright prospects for the protection or promotion of human and peoples' rights in the
continent. Three main reasons inform this brashly
pessimistic scenario. The first relates to the very
serious shortcomings of the OAU Charter on Human and
Peoples Rights. Besides that, it is even doubtful whether
the Charter will ever come into force in the.1980s given
the apparent lack of enthusiasm on the part of the majority
of African states to ratify the document. - This bleak
prognostication has not been improved by the uncertainties
that have surrounded the Pan-African Organisation since
1982. If three years after the Charter was adopted
unanimously in Nairobi only ten states have ratified the
document, which in its present form is highly inadequate
with respect to human rights protection in Africa, then
it is reasonable to expect that suggestions aimed at
strengthening it would also be flatly rejected.

Second, the worsening economic situation in all the
OAU members is not likely to enhance the promotion of
human and peoples' rights in the 1980s and beyond. Indeed,
we belive that economic stagnation and reverse growth
would only aggravate the current inequalities within
African states, with the few at the upper echelons of
the political and economic ladders getting richer while
the poor get poorer. But the poverty and squalor of the
masses of Africans would lead to demands in African
states--especially within the sub-imperial states--for
reforms in their political and economic systems, a
scenario very much akin to that in Latin America in the
late 1960s and early 1970s. As would be expected, the
consequent protests and unrest by the workers and the
African masses would lead only to further political and
economic repression and, of course, further violations of
their human rights. Arbitrary arrests of opposition and
labour leaders, as well as murder and/or mysterious "dis-
appearances" of journalists and other dissident elements
and their friends and/or relatives, would be commonplace
under such circumstances. We do not, in this type of
scenario, expect African states to pay even lip service
to what would be considered liberal democratic luxuries
such as freedom of the press, freedom of association,
and freedom to receive and disseminate information, all
of which are "guaranteed" under the present OAU human
rights charter. This scenario, then, is one in which
the machinery set up by the OAU to monitor human rights
violations in Africa--the Commission on Human and Peoples'
Rights--would be powerless to carry out its duties.

Closely linked with the worsening economic situation
in OAU member states is the whole issue of political
succession and regime stability. In that regard, we do

not anticipate an end to military coups and attempts by
the military to topple either military or civilian regimes
in Africa. In either case, though, the result would be
the same: mass arrests, arbitrary detentions and executions of dissidents and "subversives." In other words,
violations of human rights in Africa may be expected to
continue into the distant future. Our scenario takes due
regard to developments outside Africa and particularly
what we see as the resurgence of conservatism in the
capitalist world and the attendant concern with the
"Soviet threat" and its containment. The revival of cold
war politics in the West has unfortunately removed what
we consider to be a potent pressure on African leaders in
respect of the way they treat their citizens. Thus, concern with the containment of the Soviet Union by the
Reagan regime in the United States has led that country
to embrace patently repressive regimes both inside and
outside Africa; for instance, U.S. support for the
government of President Numeiry of Sudan through massive
military and economic aid. It does not portend well for
the protection and promotion of human rights in Africa
in the future.

This conclusion poses a serious challenge to the
researcher as well as to the policymaker in the future.
His/her job, then, would be to try and see how some of the
plausible negative scenarios can be averted; for instance,
how to ensure that the OAU's Human and Peoples' Rights
Charter is implemented in the future? In that regard,
we believe that future research on the OAU and/or African
foreign policies should be explicitly policy oriented.
Such policy-oriented research would assist greatly in
ensuring that some of the problems we have identified in
this conclusion are avoided, but would also help to make
the OAU much more relevant to its members and the entire
international community by the 1990s and beyond. Africa
has a future, which analysts as well as policymakers need
to envisage in comparative perspective.

Appendixes

OAU: Organisational Chart

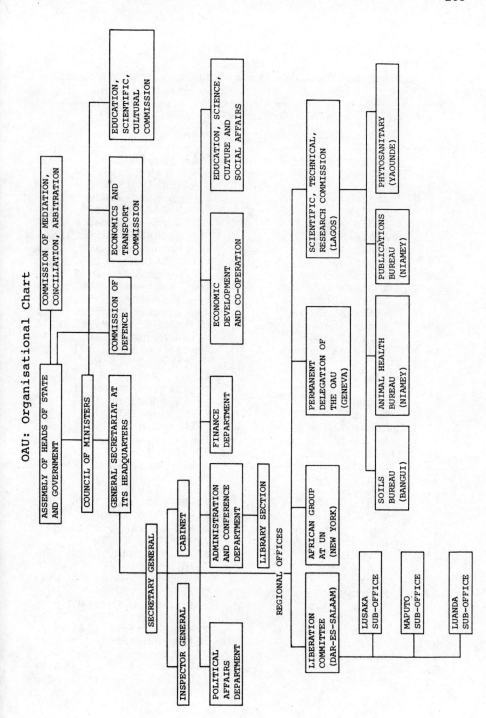

APPENDIX 2
Ordinary and Extra-Ordinary Sessions of the Council of Ministers, 1963 - 1982

Ordinary Sessions

	Venue	Date
1	Dakar	2-11 August 1963
2	Lagos	24-29 February 1964
3	Cairo	13-17 July 1964
4	Nairobi	26 February-9 March 1965
5	Accra	14-21 October 1965
6	Addis Ababa	28 February-6 March 1966
7	Addis Ababa	31 October-4 March 1966
8	Addis Ababa	27 February-4 March 1967
9	Kinshasa	4-10 September 1967
10	Addis Ababa	20-24 February 1968
11	Algiers	4-12 September 1968
12	Addis Ababa	17-22 February 1969
13	Addis Ababa	27 August-6 September 1969
14	Addis Ababa	27 February-5 March 1970
15	Addis Ababa	24-30 August 1970
16	Addis Ababa	26-28 February 1971 (then postponed)
	Addis Ababa	11-15 June 1971 (16th session resumed)
17	Addis Ababa	14-18 February 1972
18	Rabat	5-11 June 1972
19	Addis Ababa	5-10 February 1973
20	Addis Ababa	17-24 May 1973
21	Addis Ababa	27-28 February 1974
22	Mogadishu	6-12 June 1974
23	Addis Ababa	13-21 February 1975
24	Kampala	18-28 July 1975
25	Addis Ababa	23-28 February 1976
26	Port Louis	24-29 June 1976
27	Lome	21-28 February 1977
28	Libreville	23-30 June 1977
29	Tripoli	20-28 February 1978
30	Khartoum	7-16 July 1978
31	Nairobi	23 February-4 March 1979
32	Monrovia	6-16 July 1979
33	Addis Ababa	6-14 February 1980
34	Freetown	18-30 June 1980
35	Addis Ababa	23-28 February 1981
36	Nairobi	18-23 June 1981
37	Addis Ababa	22-28 February 1982
38	Tripoli I	27 July 1982 (postponed)
39	Tripoli II	15-19 November 1982 (dead-locked)

Extra-Ordinary Sessions

Venue	Date	Subject Matter
Addis Ababa	15-18 November 1963	Algerian-Moroccan conflict
Dar-es-Salaam	12-14 February 1964	Tanganyika army mutiny
Addis Ababa	5-10 September 1964	Situation in the Congo
New York, USA	16-21 December 1964	Situation in the Congo
Lagos	10-13 June 1965	Ghana-OCAM dispute
Addis Ababa	3-5 December 1965	Rhodesia
Lagos	9-11 December 1970	Portuguese invasion of Guinea
Addis Ababa	19-21 November 1973	Middle East conflict and oil crisis
Dar-es-Salaam	7-11 April 1975	Detente with South Africa
Addis Ababa	8-9 January 1976	Angolan problem
Kinshasa	6-11 December 1976	Economic issues
Lagos	20-27 April 1980	Economic issues

APPENDIX 3
Ordinary and Extra-ordinary Sessions of African Heads of State and Governments, 1965-1983

Ordinary Sessions

	Venue	Date
	Addis Ababa	22-25 May 1963 (founding conference)
1	Cairo	17-21 July 1964
2	Accra	21-26 October 1965
3	Addis Ababa	5-9 November 1966
4	Kinshasa	11-14 September 1967
5	Algiers	13-16 September 1968
6	Addis Ababa	6-9 September 1969
7	Addis Ababa	1-4 September 1970
8	Addis Ababa	21-23 June 1971
9	Rabat	12-15 June 1972
10	Addis Ababa	24-29 May 1973
11	Mogadishu	12-16 June 1974
12	Kampala	28 July-1 August 1975
13	Port Louis	2-5 July 1976
14	Libreville	2-5 July 1977
15	Khartoum	18-21 July 1978
16	Monrovia	17-21 July 1979
17	Freetown	1-4 July 1980
18	Nairobi	24-28 June 1981
19	Tripoli I and II (Non-Summit)	August and November 1982
20	Addis Ababa	8-12 June 1983

Extra-ordinary Sessions

	Venue	Date	Subject Matter
1	Addis Ababa	10-12 January 1976	Situation in Angola
2	Lagos	28-29 April 1980	Economic issues facing Africa

African Charter on Human and Peoples' Rights

PREAMBLE

The African States members of the Organization of African Unity, Parties to the present convention entitled (African Charter on Human and Peoples' Rights).

Recalling Decision 115 (XVII) of the Assembly of Heads of State and Government at its Sixteenth Ordinary Session held in Monrovia. Liberia, from 17 to 20 July 1979 on the preparation of (a preliminary draft on an African Charter on Human and Peoples' Rights providing inter alia for the establishment of bodies to promote and protect human and peoples' rights);

Considering the Charter of the Organization of African Unity, which stipulates that (freedom, equality, justice and legitimate aspirations of the African peoples);

Reaffirming the pledge they solemnly made in Article 2 of the said Charter to eradicate all forms of colonialism from Africa, to coordinate and intensify their cooperation and efforts to achieve a better life for peoples of Africa and to promote international cooperation having due regard to the Charter of the United Nations and the Universal Declaration of Human Rights;

Taking into consideration the virtues of their historical tradition and the values of African civilization which should inspire and characterize their reflection on the concept of human and peoples' rights;

Recognizing on the one hand, that fundamental human rights stem from the attributes of human beings, which justifies their international protection and on the other hand that the reality and respect of peoples' rights should necessarily guarantee human rights;

Considering that the enjoyment of rights and freedoms also implies the performance of duties on the part of everyone;

Convinced that it is henceforth essential to pay a particular attention to the right to development and that civil and political rights cannot be dissociated from economic, social and cultural rights in their conception

as well as universality and that the satisfaction of
economic, social and cultural rights is a guarantee for
the enjoyment of civil and political rights;

Conscious of their duty to achieve the total liberation of Africa, the peoples of which are still struggling
for their dignity and genuine independence, and under-
taking to eliminate colonialism, neo-colonialism, apartheid, zionism and to dismantle aggressive foreign military
bases and all forms of discrimination, language, religion
or political opinions;

Reaffirming their adherence to the principles of
human and peoples' rights and freedoms contained in the
declarations, conventions and other instruments adopted
by the Organization of African Unity, the Movement of Non-
Aligned Countries and the United Nations;

Firmly convinced of their duty to promote and
protect human and peoples' rights and freedoms taking into
account the importance traditionally attached to these
rights and freedoms in Africa:

HAVE AGREED AS FOLLOWS:

PART I

RIGHTS AND DUTIES

CHAPTER 1

HUMAN AND PEOPLES' RIGHTS

Article 1

The Member States of the Organization of African
Unity parties to the present Charter shall recognize the
rights, duties and freedoms enshrined in this Charter and
shall undertake to adopt legislative or other measures to
give effect to them.

Article 2

Every individual shall be entitled to the enjoyment
of the rights and freedoms recognized and guaranteed in
the present Charter without distinction of any kind such
as race, ethnic group, colour, sex, language, religion,
political or any other opinion, national and social
origin, fortune, birth or other states.

Article 3

1. Every individual shall be equal before the law.
2. Every individual shall be entitled to equal
   protection of the law.

Article 4

Human beings are inviolable. Every human being shall
be entitled to respect for his life and the integrity of

his person.   No one may be arbitrarily deprived of this right.

Article 5

Every individual shall have the right to the respect of the dignity inherent in a human being and to the recognition of his legal status. All forms of exploitation and degradation of man particularly slavery, slave trade, torture, cruel, inhuman or degrading punishment shall be prohibited.

Article 6

Every individual shall have the right to liberty and to the security of his person. No one may be deprived of his freedom except for reasons and conditions previously laid down by law. In particular, no one may be arbitrarily arrested or detained.

Article 7

1.   Every individual shall have the right to have his cause heard. This comprises:
    a.   The right to an appeal to competent national organs against acts of violating his fundamental rights as recognized and guaranteed by conventions, laws, regulations and customs in force.
    b.   the right to be presumed innocent until proven guilty by a competent court or tribunal;
    c.   the right to defence, including the right to be defended by counsel of his choice;
    d.   the right to be tried within a reasonable time by an impartial court or tribunal.
2.   No one may be condemned for an act or omission which did not constitute a legally punishable offense at the time it was committed. No penalty may be inflicted for an offense for which no provision was made at the time it was committed. Punishment is personal and can be imposed only on the offender.

Article 8

Freedom of conscience, the profession and free practice of religion shall be guaranteed. No one may, subject to law and order, be submitted to measures restricting the exercise of these freedoms.

Article 9

1.   Every individual shall have the right to receive information.
2.   Every individual shall have the right to express and disseminate his opinions within the law.

Article 10

1. Every individual shall have the right to free
association provided that he abides by the law.
2. Subject to the obligation of solidarity provided
for in Article 29 no one may be compelled to join an
association.

Article 11

Every individual shall have the right to assemble
freely with others. The exercise of this right shall be
subject only to necessary restrictions provided for by
law in particular those enacted in the interest of
national security, the safety, health, ethics and rights
and freedoms of others.

Article 12

1. Every individual shall have the right to freedom
of movement and residence within the borders of a State
provided he abides by the law.
2. Every individual shall have the right to leave
any country including his own, and to return to his
country. This right may only be subject to restrictions,
provided for by law for the protection of national
security, law and order, public health or morality.
3. Every individual shall have the right, when
persecuted, to seek and obtain asylum in other countries
in accordance with the law of those countries and international conventions.
4. A non-national legally admitted in a territory
of a State Party to the present Charter, may only be
expelled from it by virtue of a decision taken in
accordance with the law.
5. The mass expulsion of non-nationals shall be
prohibited. Mass expulsion shall be that which is aimed
at national, racial, ethnic or religious groups.

Article 13

1. Every citizen shall have the right to participate freely in the government of his country, either
directly or through freely chosen representatives in
accordance with the provisions of the law.
2. Every citizen shall have the right to equal
access to the public service of his country.
3. Every individual shall have the right to access
to public property and services in strict equality of
all persons before the law.

Article 14

The right to property shall be guaranteed. It may
only be encroached upon in the interest of public need or
in the general interest of the community and in accordance

with the provisions of appropriate laws.

Article 15

Every individual shall have the right to work under equitable and satisfactory conditions, and shall receive equal pay for equal work.

Article 16

1. Every individual shall have the right to enjoy the best attainable state of physical and mental health.
2. States Parties to the present Charter shall take the necessary measures to protect the health of their people and to ensure that they receive medical attention when they are sick.

Article 17

1. Every individual shall have the right to education.
2. Every individual may freely, take part in the cultural life of his community.
3. The promotion and protection of morals and traditional values recognized by the community shall be the duty of the state.

Article 18

1. The family shall be the natural unit and basis of society. It shall be protected by the State which shall take care of its physical health and morals.
2. The State shall have the duty to assist the family which is the custodian of morals and traditional values recognized by the community.
3. The State shall ensure the elimination of every discrimination against women and also ensure the protection of the rights of the woman and the child as stipulated in international declarations and conventions.
4. The aged and the disabled shall also have the right to special measures of protection in keeping with their physical or moral needs.

Article 19

All peoples shall be equal, they shall enjoy the same respect and shall have the same rights. Nothing shall justify the domination of a people by another.

Article 20

1. All peoples shall have right to existence. They shall have the unquestionable and inalienable right to self-determination. They shall freely determine their political status and shall pursue their economic and social development according to the policy they have freely chosen.

114

2. Colonized or oppressed peoples shall have the
right to free themselves from the bonds of domination by
resorting to any means recognized by the international
community.
3. All peoples shall have the right to the
assistance of the States Parties to the present Charter
in their liberation struggle against foreign domination,
be it political, economic or cultural.

Article 21

1. All peoples shall freely dispose of their wealth
and natural resources. This right shall be exercised
in the exclusive interest of the people. In no case shall
a people be deprived of it.
2. In case of spoliation the dispossessed people
shall have the right to the lawful recovery of its property as well as to an adequate compensation.
3. The free disposal of wealth and natural resources
shall be exercised without prejudice to the obligation of
promoting international economic cooperation based on
mutual respect, equitable exchange and the principles of
international law.
4. States Parties to the present Charter shall
individually and collectively exercise the right to free
disposal of their wealth and natural resources with a
view to strengthening African unity and solidarity.
5. States Parties to the present Charter shall
undertake to eliminate all forms of foreign economic
exploitation particularly that practised by international
monopolies so as to enable their peoples to fully benefit
from the advantages derived from their national resources.

Article 22

1. All peoples shall have the right to their
economic, social and cultural development with due regard
to their freedom and identity and in the equal enjoyment
of the common heritage of mankind.
2. States shall have the duty, individually or
collectively to ensure the exercise of the right to
development.

Article 23

1. All peoples shall have the right to national and
international peace and security. The principles of
solidarity and friendly relations implicitly affirmed by
the Charter of the United Nations and reaffirmed by that
of the Organization of African Unity shall govern
relations between States.
2. For the purpose of strengthening peace,
solidarity and friendly relations, States parties to the
present Charter shall ensure that:

a. any individual enjoying the right of
    asylum under Article 12 of the present
    Charter shall not engage in subversive
    activities against his country of origin
    or any other State party to the present
    Charter;
b. their territories shall not be used as
    bases for subversive or terrorist
    activities against the people of any
    other State party to the present Charter.

Article 24

All peoples shall have the right to a general satisfactory environment favourable to their development.

Article 25

States parties to the present Charter shall have the
duty to promote and ensure through teaching, education
and publication, the respect of the rights and freedoms
contained in the present Charter and to see to it that
these freedoms and rights as well as corresponding
obligations and duties are understood.

Article 26

States parties to the present Charter shall have the
duty to guarantee the independence of the Courts and shall
allow the establishment and improvement of appropriate
national institutions entrusted with the promotion and
protection of the rights and freedoms guaranteed by the
present Charter.

CHAPTER II

DUTIES

Article 27

1. Every individual shall have duties towards his
family and society, the State and other legally recognised communities and the international community.
2. The rights and freedoms of each individual shall
be exercised with due regard to the rights of others,
collective security, morality and common interest.

Article 28

Every individual shall have the duty to respect and
consider his fellow beings without discrimination, and to
maintain relations aimed at promoting, safeguarding and
reinforcing mutual respect and tolerance.

Article 29

The individual shall also have the duty:

1. To preserve the harmonious development of the
    family and to work for the cohesion and respect
    of the family; to respect his parents at all
    times, to maintain them in case of need.
2. To serve his national community by placing his
    physical and intellectual abilities at its
    service.
3. Not to compromise the security of the State
    whose national or resident he is;
4. To preserve and strengthen social and national
    solidarity, particularly when the latter is
    threatened;
5. To preserve and strengthen the national indepen-
    dence and the territorial integrity of his
    country and to contribute to its defence in
    accordance with the law;
6. To work to the best of his abilities and
    competence, and to pay taxes imposed by law in
    the interest of the society;
7. To preserve and strengthen positive African
    cultural values in his relations with other
    members of the society, in the spirit of
    tolerance, dialogue and consultation and, in
    general, to contribute to the promotion of the
    moral well being of society.
8. To contribute to the best of his abilities, at
    all times and at all levels, to the promotion
    and achievement of African unity.

PART II

MEASURES OF SAFEGUARD

CHAPTER I

ESTABLISHMENT AND ORGANISATION OF THE AFRICAN COMMISSION ON HUMAN AND PEOPLES' RIGHTS

Article 30

An African Commission on Human and Peoples' Rights,
hereinafter called (the Commission), shall be established
within the Organisation of African Unity to promote human
and peoples' rights and ensure their protection in Africa.

Article 31

1. The commission shall consist of eleven members
chosen from amongst African personalities of the highest
reputation, known for their high morality, integrity,
impartiality and competence in matters of human and
peoples' rights; particular consideration being given
to persons having legal experience.

2.   The members of the Commission shall serve in their personal capacity.

Article 32

The Commission shall not include more than one national of the same State.

Article 33

The members of the Commission shall be elected by secret ballot by the Assembly of Heads of State and Government, from a list of persons nominated by the States parties to the present Charter.

Article 34

Each State party to the present Charter may not nominate more than two candidates. The candidates must have the nationality of one of the States parties to the present Charter. When two candidates are nominated by a State, one of them may not be a national of that State.

Article 35

1.   The Secretary General of the Organization of African Unity shall invite States parties to the present Charter at least four months before the elections to nominate candidates;
2.   The Secretary General of the Organization of African Unity shall make an alphabetical list of the persons thus nominated and communicate it to the Heads of State and Government at least one month before the elections.

Article 36

The members of the Commission shall be elected for a six year period and shall be eligible for re-election. However, the term of office of four of the members elected at the first election shall terminate after two years and the term of office of three others, at the end of four years.

Article 37

Immediately after the first election, the Chairman of the Assembly of Heads of State and Government of the Organization of African Unity shall draw lots to decide the names of those members referred to in Article 36.

Article 38

After their election, the members of the Commission shall make a solemn declaration to discharge their duties impartially and faithfully.

Article 39

1. In case of death or resignation of a member of the Commission the Chairman of the Commission shall immediately inform the Secretary General of the Organization of African Unity, who shall declare the seat vacant from the date of death or from the date on which the resignation takes effect.
2. If, in the unanimous opinion of other members of the Commission, a member has stopped discharging his duties for any reason other than a temporary absence, the Chairman of the Commission shall inform the Secretary General of the Organization of African Unity, who shall then declare the seat vacant.
3. In each of the cases anticipated above, the Assembly of Heads of States and Government shall replace the member whose seat became vacant for the remaining period of his term unless the period is less than six months.

Article 40

Every member of the Commission shall be in office until the date his successor assumes office.

Article 41

The Secretary General of the Organization of African Unity shall appoint the Secretary of the Commission. He shall provide the staff and services necessary for the effective discharge of the duties of the Commission. The Organization of African Unity shall bear cost of the staff and services.

Article 42

1. The Commission shall elect its Chairman and Vice Chairman for a two-year period. They shall be eligible for re-election.
2. The Commission shall lay down its rules of procedure.
3. Seven members shall form the quorum.
4. In case of an equality of votes, the Chairman shall have a casting vote.
5. The Secretary General may attend the meetings of the Commission. He shall neither participate in deliberations nor shall he be entitled to vote. The Chairman of the Commission may, however, invite him to speak.

Article 43

In discharging their duties, members of the Commission shall enjoy diplomatic privileges and immunities provided for in the General Convention on the Privileges and immunities of the Organisation of African Unity.

Article 44

Provision shall be made for the emoluments and allowances of the members of the Commission in the Regular Budget of the Organization of African Unity.

CHAPTER II

MANDATE OF THE COMMISSION

Article 45

The functions of the Commission shall be:

1. To promote Human and Peoples' Rights and in particular
   a. To collect documents, undertake studies and researches on African problems in the field of human and peoples' rights, organize seminars, symposia and conferences, disseminate information, encourage national and local institutions concerned with human and peoples' rights, and should the case arise give its views or make recommendations to Governments.
   b. to formulate and lay down, principles and rules aimed at solving legal problems relating to human and peoples' rights and fundamental freedom upon which African Governments may base their legislation.
   c. Co-operate with other African and international institutions concerned with the promotion and protection of human and peoples' rights.
2. Ensure the protection of human and peoples' rights under conditions laid down by the present Charter.
3. Interpret all the provisions of the present Charter at the request of a state Party, an institution of the OAU or an African Organization recognized by the OAU.
4. Perform any other tasks which may be entrusted to it by the Assembly of Heads of State and Government.

CHAPTER III

PROCEDURE OF THE COMMISSION

Article 46

The commission may resort to any appropriate method of investigation; it may hear from the Secretary General of the Organization of African Unity or any other person capable of enlightening it.

COMMUNICATION FROM STATES

Article 47

If a state party to the present Charter has good reasons to believe that another State Party to this Charter has violated the provision of the Charter, it may draw, by written communication, the attention of that State to the matter. This communication shall also be addressed to the Secretary General of the OAU and to the Chairman of the Commission. Within three months of the receipt of the communication, the State to which the communication is addressed shall give the enquiring State, written explanation or statement elucidating the matter. This should include as much as possible relevant information relating to the laws and rules of procedure applied and applicable and the redress already given or course of action available.

Article 48

If within three months from the date on which the original communication is received by the State to which it is addressed, the issue is not settled to the satisfaction of the two States involved through bilateral negotiation or by any other peaceful procedure, either State shall have the right to submit the matter to the Commission through the Chairman and shall notify the other States involved.

Article 49

Notwithstanding the provisions of Article 47, if a State party to the present Charter considers that another State party has violated the provisions of the Charter, it may refer the matter directly to the Commission by addressing a communication to the Chairman, to the Secretary General of the Organization of African Unity and the State concerned.

Article 50

The Commission can only deal with a matter submitted to it after making sure that all local remedies, if they exist, have been exhausted, unless it is obvious to the Commission that the procedure of achieving these remedies would be unduly prolonged.

Article 51

1. The Commission may ask the States concerned to provide it with all relevant information.
2. When the Commission is considering the matter, States concerned may be represented before it and submit written or oral representation.

Article 52

After having obtained from the States concerned and from other sources all the information it deems necessary and after having tried all appropriate means to reach an amicable solution based on the respect of Human and Peoples' Rights, the Commission shall prepare, within a reasonable period of time from the notification referred to in Article 48, a report to the State concerned and communicated to the Assembly of Heads of State and Government.

Article 53

While transmitting its report, the Commission may make to the Assembly of Heads of State and Government such recommendations as it deems useful.

Article 54

The Commission shall submit to each ordinary Session of the Assembly of Heads of State and Government a report on its activities.

OTHER COMMUNICATIONS

Article 55

1.   Before each Session, the Secretary of the Commission shall make a list of the Communications other than those of States parties to the present Charter and transmit them to the Members of the Commission, who shall indicate which communications should be considered by the commission.

2.   A communication shall be considered by the Commission if a simple majority of its members so decide.

Article 56

Communication relating to human and peoples' rights referred to in Article 55 received by the commission, shall be considered if they:

1.   indicate their authors even if the latter request anonymity.
2.   are compatible with the Charter of the Organisation of African Unity or with the present Charter.
3.   are not written in disparaging or insulting language directed against the State concerned and its institutions or to the Organisation of African Unity.
4.   are not based exclusively on news disseminated through the mass media.
5.   are sent after exhausting local remedies, if any unless it is obvious that this procedure is unduly prolonged.

6. are submitted within a reasonable period from
    the time local remedies are exhausted or from
    the date the commission is seized with the
    matter and
7. do not deal with cases which have been settled
    by these states involved in accordance with
    the principles of the Charter of the United
    Nations, or the Charter of the Organization of
    African Unity or the provisions of the present
    Charter.

Article 57

Prior to any substantive consideration, all communications shall be brought to the knowledge of the State concerned by the Chairman of the Commission.

Article 58

1. When it appears after deliberations of the
Commission that one or more communications apparently
relate to special cases which reveal the existence of a
series of serious or massive violations of human and
peoples' rights, the Commission shall draw the attention
of the Assembly of Heads of State and Government to these
special cases.
2. The Assembly of Heads of State and Government
may then request the Commission to undertake an in-depth
study of these cases and make a factual report,
accompanied by its finding and recommendations.
3. A case of emergency duly noticed by the
Commission shall be submitted by the latter to the
Chairman of the Assembly of Heads of State and Government
who may request an in-depth study.

Article 59

1. All measures taken within the provisions of the
present Chapter shall remain confidential until such a
time as the Assembly of Heads of State and Government
shall otherwise decide.
2. However, the report shall be published by the
Chairman of the Commission upon the decision of the
Assembly of Heads of State and Government.
3. The report on the activities of the Commission
shall be published by its Chairman after it has been
considered by the Assembly of Heads of State and
Government.

CHAPTER IV

APPLICABLE PRINCIPLES

Article 60

The Commission shall draw inspiration from

international law on human and peoples' rights, particularly from the provisions of various African instruments
on human and peoples' rights, the Charter of the United
Nations, the Charter of the Organization of African Unity,
the Universal Declaration of Human Rights, other instruments adopted by the United Nations and by African
countries in the field of human and peoples' rights as
well as from the provisions of various instruments adopted
within the Specialised Agencies of the United Nations of
which the parties to the present Charter are members.

Article 61

The Commission shall also take into consideration,
as subsidiary measures to determine the principles of law,
other general or special international conventions, lay-
ing down rules expressly recognized by member states of
the Organization of African Unity, African practices
consistent with international norms on human and peoples'
rights, customs generally accepted as law, generally
principles of law recognized by African states as well as
legal precedents and doctrine.

Article 62

Each State party shall undertake to submit every two
years, from the date the present Charter comes into force,
a report on the legislative or other measures taken with a
view to giving effect to the rights and freedoms recognized and guaranteed by the present Charter.

Article 63

1. The present Charter shall be open to signature,
ratification or adherence of the member states of the
Organization of African Unity.
2. The instruments of ratification or adherence to
the present Charter shall be deposited with the Secretary
General of the Organization of African Unity.
3. The present Charter shall come into force three
months after the reception by the Secretary General of
the instruments of ratification or adherence of a simple
majority of the member states of the Organization of
African Unity.

PART III

GENERAL PROVISIONS

Article 64

1. After the coming into force of the present
Charter, members of the Commission shall be elected in
accordance with the relevant Articles of the present
Charter.

2. The Secretary General of the Organization of African Unity shall convene the first meeting of Commission at the Headquarters of the Organization within three months of the constitution of the Commission. Thereafter, the Commission shall be convened by its Chairman whenever necessary but at least once a year.

Article 65

For each of the States that will ratify or adhere to the present Charter after its coming into force, the Charter shall take effect three months after the date of the deposit by that State of its instrument of ratification or adherence.

Article 66

Special protocols or agreements may, if necessary, supplement the provisions of the present Charter.

Article 67

The Secretary General of the Organization of African Unity shall inform member states of the Organization of the deposit of each instrument of ratification or adherence.

Article 68

The present Charter may be amended if a State party makes a written request to that effect to the Secretary General of the Organization of African Unity. The Assembly of Heads of State and Government may only consider the draft amendment after all the States parties have been duly informed of it and the Commission has given its opinion on it at the request of the sponsoring State. The amendment shall be approved by a simple majority of the States parties. It shall come into force for each State which has accepted it in accordance with its constitutional procedure three months after the Secretary General has received notice of the acceptance.

Abbreviations

AHG	Assembly of African Heads of State and Governments
ALC	African Liberation Committee
ANC	African National Congress
CIAS	Conference of Independent African States
CM	Council of Ministers
ECA	Economic Commission for Africa
ECM	Emergency Council of Ministers
EEC	European Economic Community
FAO	Food and Agricultural Organisation
FNLA	National Front for the Liberation of Angola
FROLINAT	Front for the Liberation of Chad
GRAE	Government of Angola in Exile
GUNT	Transitional Government of National Unity
ICJ	International Court of Justice
IMF	International Monetary Fund
LDC	Less Developed Countries
MOJA	Movement for Justice in Africa
MPLA	Popular Movement for the Liberation of Angola
NIEO	New International Economic Order
OAU	Organisation of African Unity
OPEC	Organisation of Petroleum Exporting Countries
PAC	Pan-African Congress
PAIGC	African Party for the Independence of Guinea and Cape Verde
PF	Patriotic Front

POLISARIO	Popular Front for the Liberation of Western Sahara
SADR	Saharan Arab Democratic Republic
SWAPO	South West Africa Peoples Organisation
UDI	Unilateral Declaration of Independence
UN	United Nations
UNITA	Union for the Total Independence of Angola
ZANU	Zimbabwe African National Union
ZAPU	Zimbabwe African Peoples Union

Index